DETERIORATION AND OPTIMAL REHABILITATION MODELLING FOR URBAN WATER DISTRIBUTION SYSTEMS

Yi ZHOU

DETERIORATION AND OPTIMAL REHABILITATION MODELLING FOR URBAN WATER DISTRIBUTION SYSTEMS

DISSERTATION

Submitted in fulfillment of the requirements of
the Board for Doctorates of Delft University of Technology
and
of the Academic Board of the IHE Delft
Institute for Water Education
for
the Degree of DOCTOR
to be defended in public on
Monday, 7 May 2018, at 12.30 hours
in Delft, the Netherlands

by

Yi ZHOU
Master of Science in Environment Engineering, Wuhan University
born in Hubei, China

This dissertation has been approved by the
promotor: Prof. dr.   K. Vairavamoorthy

Composition of the doctoral committee:

Chairman   Rector Magnificus TU Delft
Vice-Chairman   Rector IHE Delft
Prof. dr. K. Vairavamoorthy   IHE Delft / TU Delft, promotor

Independent members:
Prof.dr.ir. L.C. Rietveld   TU Delft
Prof.dr.ir. C. Zevenbergen   IHE Delft / TU Delft
Prof.dr. S. Mohan   Indian Institute of Technology, Madras, India
Prof.dr. J. Xia   Wuhan University, China
Prof.dr. ir. A. E. Mynett   TU Delft/ IHE Delft, reserve member

CRC Press/Balkema is an imprint of the Taylor & Francis Group, an informal business

Published by:
CRC Press/Balkema
PO Box 11320, 2301 EH Leiden, The Netherlands
Pub.NL@taylorandfrancis.com
www.crcpress.com – www.taylorandfrancis.com
ISBN 978-1-138-32281-3

Abstract

Water distribution systems are a major component of a water utility's asset and may constitute over half of the overall cost of a water supply system. They are critical in delivering water to consumers from a variety of sources. Pipe failures within the distribution system can have a serious impact to both people's daily life and to the wastage of limited, high quality water that has undergone extensive treatment. Hence it is important to maintain the condition and integrity of distribution systems.

This thesis presents a whole-life cost optimisation model for the rehabilitation of water distribution systems. This model allows decision makers to prioritize their rehabilitation strategy in a proactive and cost-effective manner. The optimisation model presented in this thesis, combines a pipe breakage number prediction model with a pipe criticality assessment model, that enables the creation of a well-constructed and more tightly constrained optimisation model. This results in improved convergence and reduced computational time and effort. The resulting optimisation model is a multiple-objective one that is solved using an improved genetic algorithm technique.

The first model developed is a pipe breakage number prediction model. This model combines information on the physical characteristics of the pipes (i.e., pipe age, diameter, length, material etc.) with historical information on breakage and failure rates. It uses this information to group pipes based on their condition and general deterioration tendency. A weighted multiple nonlinear regression analysis is applied to develop a model describing the condition of different the pipe groups.

The second model is a criticality assessment model. This model combines a pipe's condition with its hydraulic significance (i.e. how important a pipe is to hydraulic performance of the network), to establish its criticality to the network. The criticality index is calculated using a multi-criteria decision making method, the Technique for Order of Preference by Similarity to Ideal Solution (TOPSIS). The thesis applies a modified TOPSIS approach that avoids the problem that the rank calculation method is inconsistent with the TOPSIS principle, common in traditional TOPSIS. The application of a pipe criticality assessment model enables the preliminarily screening of pipes in a water distribution system and allows the optimisation model to focus its efforts on those pipes that are most important for a rehabilitation strategy, improving convergence and computational performance.

The third model developed, is a whole life cost optimal rehabilitation model. It is a multiple-objective and multiple-stage model. The objectives of the model include whole-life cost minimization and benefit maximization. Benefit is articulated in terms of burst number minimization and hydraulic reliability maximisation. The objectives are optimized subject to financial and hydraulic performance constraints. The optimisation model is solved using genetic algorithms, namely a modified NSGA-II. The modifications applied to the NSGA-II, includes an induced mutation process that improves the search process. The application of the optimisation model, provides decision makers with a suite of rehabilitation decisions that minimise the whole life cost of the network while maximising its long-term performance.

To demonstrate the efficacy of the developed models, the thesis includes their application to case-study networks. The results are described and discussed in detail in terms their utility from the perspective of a decision maker. It is envisaged that the developed modelling tools will be used by water utilities to improve their decision making process in relation to pipe rehabilitation and more generally asset management.

Samenvatting

Waterdistributiesystemen zijn een belangrijk onderdeel van het bezit van een drinkwaterbedrijf en kunnen meer dan de helft van de totale kosten van een watervoorzieningssysteem uitmaken. Ze zijn van cruciaal belang voor het leveren van water aan consumenten uit verschillende bronnen. Pijpstoringen in het distributiesysteem kunnen een ernstige impact hebben op het dagelijks leven van mensen en op de verspilling van een water van hoge kwaliteit dat een uitgebreide behandeling heeft ondergaan en slechts beperkt beschikbaar is. Daarom is het belangrijk om de staat en de volledigheid van distributiesystemen te behouden.

Dit proefschrift presenteert een kosten-optimalisatiemodel voor het herstel van waterdistributiesystemen gedurende de hele levensduur (life-cycle costing). Met dit model kunnen besluitvormers hun strategie voor herstelwerkzaamheden op een proactieve en kosteneffectieve manier prioriteren. Het optimalisatiemodel dat in dit proefschrift wordt gepresenteerd, combineert een model dat de conditie van pijpleidingen beoordeeld met een kritisch beoordelingsmodel, waardoor een goed geconstrueerd en efficiënter optimalisatiemodel kan worden gecreëerd. Dit resulteert in verbeterde convergentie en verminderde computertijd en -vermogen. Het resulterende optimalisatiemodel is een multi-objectief model dat wordt opgelost met behulp van een verbeterde genetische algoritme-techniek.

Het eerste ontwikkelde model is een voorspellingsmodel voor het aantal leidingbreuken. Dit model combineert informatie over de fysieke kenmerken van de buizen (d.w.z. de leeftijd van de pijp, diameter, lengte, materiaal, enz.) met historische informatie over de verhouding van het aantal falen en breuken. Het gebruikt deze informatie om leidingen te groeperen op basis van hun toestand en algemene neiging tot verslechtering. Een gewogen meervoudige niet-lineaire regressieanalyse wordt toegepast om een model te ontwikkelen dat de toestand van verschillende pijpgroepen beschrijft.

Het tweede model is een kritikaliteitsbeoordelingsmodel. Dit model combineert de conditie van een pijpleiding met zijn hydraulische significantie (dat wil zeggen, hoe belangrijk een leiding is voor de hydraulische prestaties van het netwerk), om vast te stellen hoe krititiek deze is voor het netwerk. Een index die het kritisch karakter aangeeft wordt berekend met behulp van een beslissingsmethode gebaseerd op meerdere criteria, de Technique for Order of Preference by Similarity to Ideal Solution (TOPSIS). Het proefschrift past een aangepaste TOPSIS-benadering toe die het probleem vermijdt dat de rangberekeningsmethode niet strookt met het TOPSIS-principe, dat gebruikelijk is in traditionele TOPSIS. De toepassing van een kritikaliteitsbeoordelingsmodel voor pijpleidingen maakt de preliminaire screening van leidingen in een waterdistributiesysteem mogelijk en stelt het optimalisatiemodel in staat om inspanningen te concentreren op die buizen die het belangrijkst zijn voor een herstelstrategie, waardoor convergentie en computationele prestaties worden verbeterd.

Het derde model dat is ontwikkeld, is een herstelkosten-optimalisatiemodel toegepast op de hele levensduur (life-cycle costing). Het is een model met meerdere doelen en meerdere fasen. De doelstellingen van het model omvatten de kostenminimalisatie voor de gehele levenscyclus en maximalisatie van de voordelen. Het voordeel wordt uitgedrukt in termen van het aantal minimale leidingbreuken en de maximale hydraulische betrouwbaarheid. De doelstellingen zijn geoptimaliseerd voor financiële en hydraulische prestatiebeperkingen. Het optimalisatiemodel wordt bepaald met behulp van genetische algoritmen, namelijk een gemodificeerde NSGA-II. De modificaties toegepast op de NSGA-II omvatten een geïnduceerd mutatieproces dat het zoekproces verbetert. De toepassing van het optimalisatiemodel biedt besluitvormers een reeks herstelmogelijkheden die de kosten tijdens de hele levensduur van het netwerk minimaliseren en tegelijkertijd de prestaties op de lange termijn maximaliseren.

Om de doeltreffendheid van de ontwikkelde modellen te demonstreren, beschrijft het proefschrift hoe deze zijn toegepast op netwerken in verschillende case studies. De resultaten worden beschreven en uitvoerig besproken in termen van hun nut vanuit het perspectief van een besluitvormer. Het is de bedoeling dat de ontwikkelde modelleringsinstrumenten zullen worden gebruikt door waterbedrijven om hun besluitvormingsproces met betrekking tot herstel van pijpleidingen en meer in het algemeen vermogensbeheer te verbeteren.

Table of Contents

Chapter 1 Introduction

1.1 Background

Water distribution systems (WDS) are considered a critical component of an urban infrastructure system. The problem of aging and deterioration is an inevitable and natural tendency of WDS infrastructures, although they are well designed, carefully protected and operated. Such deterioration issue is a growing concern to WDS managers. Water mains reach the end of its service life gradually. An obvious example is the growing number of burst pipes in some areas. The deterioration of WDS cause many negative effects to water utility and customers, such as an increase in the number of breakages, leakages, roughness growing and water quality deterioration. For customers, this results in a reduction of the quality of the service, whilst for water utilities, such deterioration results in operation and maintenance costs.

A WDS is one of the most expensive components of an urban water supply system. The maintenance and operation costs incurred to combat its deterioration are expensive. Developing urbanization and a large population need more infrastructure and capital input to support the normal function of a WDS. The aging tendency of WDS infrastructure is also accelerated due to climate change and increasing costs of assets.

Since a water distribution network is a large scale and interrelated system, pipe replacement and other maintenance actions will result in far-reaching and complex consequences to the system, instead of being limited to a separate pipes or local customers. Therefore, the rehabilitation of water mains is an important part of effectively managing a WDS. It is a technological and economical challenge for water service utilities as well. As such, a long-term and global rehabilitation strategy is necessary.

1.1.1 Water Distribution System (WDS) Deterioration Issues

The most typical and direct phenomena of urban water distribution system decay are pipeline damage, pipe bursts and water leakages. The accompanying performances include irregular water supply, insufficient water pressure, water quality pollution, reliability decline of pipe network etc. The consequences of an urban WDS decay are mainly as follows:

(1) Immediate consequences - Water supply utilities have to face the loss of water leakages, the increase in maintenance, and the increased cost of pipe network updating, which eventually lead to an increase in the total operating costs. For users, the water pressure and regularity of water supply are affected.

(2) Indirect consequences - The damage of the pipeline causes the possibility of infiltration of pollutants into the pipe, losses caused to third parties, e.g. house foundations being damaged by soaking, the waste of water resources, and the decline of the reputation of the company etc.

There are some difficulties in solving the deterioration problem:

1. Huge System

The water distribution system of a city usually covers the entire city, is a complex system and accumulates a large amount of assets. A water supply network system is developed over the years in keeping with the development of a modern city.

2. Complicated System

A WDS is usually composed of pumps, pipes, water tanks etc., forming complex assets. In management, multiple attributes (pipe material, length, installation time, location, explosion, maintenance, replacement records etc.) of each object have to be recorded, forming a large amount of complex asset data. Owing to the system of the network distribution, local changes have different effects on the system's performance. The maintenance, renewal and natural aging process exist simultaneously, where the aging process is slow but ongoing. There are many factors that affect the network's decline, and the mechanism is complex.

3. Insufficient Data

Historical data and observational data help to analyse the general law of pipe deterioration. However, the accumulation of historical data is usually insufficient. Meanwhile, the existing observation methods are limited, and the cost of data acquisition is very high. Data shortage causes more difficulties.

4. Uncertainty of Development

The degree of aging and deterioration of pipes, as well as the development of water consumption, is accompanied with obvious randomness. Since future development and the current forecast are inevitably biased, the decision for updating, based on the predicted situation, is difficult to guarantee when the development and prediction are inconsistent.

1.1.2 Water Distribution System Deterioration Process

The deterioration of a WDS affects various performances of the system and results in declined service levels to customers, such as leakage, declined hydraulic performance, water quality degradation, water supply interruption etc. As a result of deterioration, pipe structural integrity is destroyed and the resistance capability to the environment and operation stress declines gradually. When the stress on the pipe exceeds its resistance capability, pipe breakage or failure is likely to occur. As pipe deterioration is one of the main causes of various systematic failures, and brings on declined performances or even risks, it is a major concern to asset managers and decision makers.

Pipe breakage is one of the most obvious results of pipe deterioration. Many pipe failure events which are the result of various deterioration mechanisms (e.g. internal/external corrosion or surface loads etc.), are described using the term pipe breakage. These include: pipe body cracks or splits; joint failures; and hydrant valve failures. These types of failures are often detected by the operators of the network and repair records are maintained.

Common terminology refers to water main bursts as breaks, breakages or failures (Farley and Trow 2003). The terms of pipe failure, break and burst often have the same implication in literatures.

For a typical single pipe, it is thought that its service reliability decreases with time (Figure 1.1). This can be divided into five steps: (1) installation, (2) initiation of corrosion, (3) crack before leak, (4) partial failure, and (5) complete failure. Such a curve indicates a gradual deterioration process.

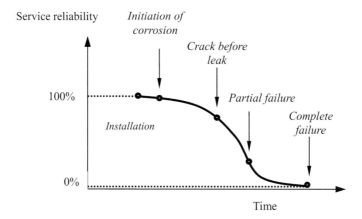

Figure 1.1   Pipe failure development (after Misiunas (2005))

Viewed from pipe group's aspect, pipe failure rate with time can be described as "Bathtub Curve" (Figure 1.2), which is widely used in reliability engineering. It indicates the whole network's general failure rate changing with time. A particular form of the failure function comprises of three parts:

(1) The first part is "*Early Failure Period,*" which has a decreasing failure rate, known as early failures. This phase is the period right after installation, in which breaks occur mainly as a result of poor production and poor workmanship during installation. The failure rate is high but quickly decreases as defective products are identified and discarded, and early sources of potential failure, such as manufacture flaw, handling and installation error appear.

(2) The second part is "*Intrinsic Failure Period,*" which has a relatively constant failure rate, known as random failures. Pipe operates relatively trouble free, with some low failure frequency resulting from random phenomena such as random heavy loads, third party interference, etc.

(3) The third part is "*Wear-out Failure Period,*" which has an increasing failure rate, known as wear-out failures. This is a period of increasing failure frequency due to deterioration of the pipe material (e.g. corrosion) which finally leads to the collapse of the pipe.

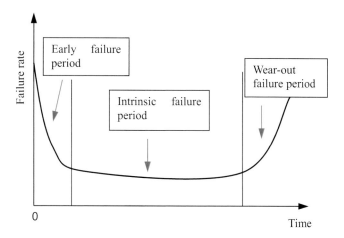

Figure 1.2   Bathtub curve

The bathtub curve denotes a statistic rule for pipe failure. The curve trend in the graph presents the general failure rate tendency with time for the whole pipe population. Failure rate or failure probability is relatively low. Therefore, the trend cannot be observed through one pipe or a few pipes. The time span of each phase may vary dramatically for various pipes under various conditions. The deterioration mechanism differs in each phase.

In the early failure period, the main inherent failure cause is manufacture flaw or poor installation quality. In the intrinsic failure period, the low failure rate is because the bad pipes have been purged and the remainder pipes are strong enough to resist various stresses. This is like a middle aged person who has low probability to get ill. The third one is the wearing out failure period, which can be thought that pipes become weak after long service and there is higher probability of failure.

To show all three stages, complete failure data is necessary. These data usually go back to when and where the pipes were laid, its repair history, or even the surrounding conditions. To obtain this complete data is difficult or even practically impossible.

The "bathtub curve" in Figure 1.2 can describe the failure probability for a group of pipes during their whole service time without any pipe rehabilitation. If the components (e.g., pipes in a distribution system) can be replaced or renewed, the whole system's failure rate will not be the "bathtub curve" in the graph. Because the entire system's performance depends on the group pipes' performance (e.g., failure rate) instead of any individual pipe's, the failure rate can be always kept at a low level if well maintained.

Since water distribution system has always been in existence as long as the city, its maintenance, renewal and rehabilitation must be viewed in the long term. The costs have to be considered in the long term and from a more comprehensive view. Whole-life cost (WLC), or Life-cycle cost (LCC), refers to the cost of an asset over its entire life (i.e. total cost of ownership). It includes both economic costs that are relatively easy to quantify, as well as environmental costs and social costs that are less easy to quantify. When calculating WLC, we need to include a diverse range of expenditure such as costs involved with project planning, design and construction, operation and maintenance, renewal and rehabilitation, depreciation and disposal.

The deterioration of WDS is a general tendency, and all kinds of maintenance, renewal and rehabilitation actions are the contrary efforts to delay or relieve the deterioration. The performance of a WDS is determined by the two conflict tendency. How much effort can be made to combat the deterioration is determined by the total cost, especially the whole life cost.

1.2 Pipe Deterioration, Failure and Rehabilitation

1.2.1 Definition of Failure

As soon as a pipe is installed underground, its deterioration process starts. Pipe material aging is coupled with the continuous and discontinuous stress placed on these systems by operational and environmental conditions. When the residual stress resistance cannot sustain internal or external stress, pipe break will occur. Pipe structural deterioration and one of its consequences, structural failure (i.e., "break" or "burst"), is easy to define and identify. Therefore, pipe breakage rate for a group of pipes and pipe condition assessment for an individual pipe are applied to address the pipe deterioration degree. Pipe structural deterioration is typical but not the only type of pipe deterioration. The deterioration of pipe can be classified into two categories: (1) Structural deterioration, which diminishes pipe's residual structural resistance capacity; and (2) Non-structural deterioration, characterized by more roughness of pipe inner surfaces and narrower inner diameter, resulting in diminishing hydraulic capacity, degradation of water quality and even residual structural resistance in case of severe corrosion.

A failure in a WDS can be defined in many ways, including: a reduction in service pressure, below a specified minimum; an unplanned interruption to supply; an event that leads to an (negative) impact on the physical, chemical or biological quality of the water etc. Hence, before undertaking a probability analysis of failures in a WDS, it is important to be clear on what is meant by a failure. With the deterioration categories, pipe failure can be classified into three categories (Rajani and Kleiner 2002):

1.   Structural Failure

Physical rupture of a water main is fairly easy to define, i.e., "break" or "burst" failure, where an active repair intervention is required. Pipe structural failure play a dominant role in various failure performances and, therefore, it is the focus of this study. For example, when one pipe is broken, this makes the transported water exposed to the external environment which has potential contamination risk. The lost water from the breakage point also reduces the hydraulic capacity. Of most research concern is pipe exterior deterioration, as it is the principal contributor to structural failure of pipes.

2.   Hydraulic Failure

A hydraulic failure is usually defined as the inability of the WDS to meet the water demand with a specified minimum pressure. As described by Rajani and Kleiner (2002), a hydraulic failure can occur for many reasons including: demand in the system being greater than anticipated (this could be due to heavy leaks); a physical component failure (e.g., pipe burst or pump shutdown); and severe deterioration in the condition of the pipes resulting in a reduction in the carrying capacity of the network.

3.   Water Quality Failure

The interactions of delivered water and pipe material may lead to complicated chemical and bio-chemical reactions in the system. Pipe corrosion is a typical case. If external impurities intrude into the system through some breaks, the reaction process will be more complex. Water quality failures may result in changes to the physical, chemical or biological characteristics of the water. Water quality failures are often classified based on the way in which the failure occurred: (1) ingress and intrusion of contaminants through leaks and cracks in the pipe; (2) bacteria regrowth along the pipe walls; (3) leaching of chemicals and corrosion products from the pipe walls; (4) permeation of organic products from components such as gaskets, into the water. It is the first class of failure that relates to pipe structural issues.

Pipe deterioration reduces a system's reliability and increases its potential vulnerability. Among these deterioration categories, structural deterioration plays a dominate role and has a close relationship with the other two deterioration categories. Therefore, pipe structural deterioration, or structural failure, has been the major research topic of pipe deterioration.

1.2.2 Failure Risk in a Water Distribution Network

The deterioration of pipe is usually characterized by increasing pipe breakage water leakage or changes in internal wall roughness. Major leakage is usually characterized by pipe breaks or pipe bursts, and this type of leakage will affect the flow of water in the pipeline, resulting in pressure losses that affect the hydraulic performance. In addition, it is through leaks that contaminants enter the pipes and hence leakage (and frequent pipe breaks), can have a detrimental effect on water quality. Internal pipe deterioration will affect the carrying capacity of the pipe and hence impact the hydraulic performance. Internal tuberculation can also become sites for bacterial regrowth and so can negatively impact water quality. Incremental but continuous pipe deterioration will ultimately result in pipe failure and brings on more failure risk. Figure 1.3 shows a general framework for comprehensive decision making for WDS renewal.

In the context of reliability engineering and risk management, one of the risk definitions depends on the type of asset or system. For buried pipes, it can be defined that the failure risk is the mathematical expectation of the failure consequence, i.e,

Failure risk= failure probability × failure consequence (costs)

Apparently, it is easy to understand and calculate the risk through the above formula. However, the diversity of failure definition and the diversity of consequence, including that of measurements, and failure probability estimation, often make risk assessment difficult as well. Take the general total costs of failure as an example: it includes direct cost, indirect cost and social costs (Skipworth, *et al.* 2002). However, there is no universal measurement to quantify indirect and social costs due to the difficulties described previously. Moreover, the definitions of failure are so diversified and failure probability is difficult to describe. Therefore, accurate quantification of risk of failure is difficult and almost impossible, even though it is certain that the frequency/probability of all of the three kinds of failure always increases with deterioration. Failures on different pipes usually result in different costs.

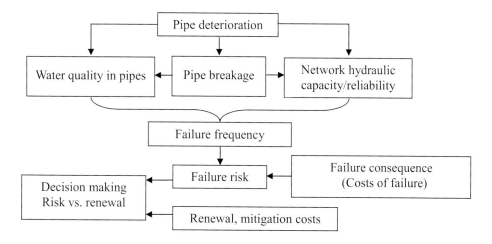

Figure 1.3   A general framework for decision making in water distribution system (after (Rajani and Kleiner 2002))

This product function about risk is only a conceptual formula which is not always applied directly in most practical calculations. Hence, a new measurement system which quantifies a pipe's failure risk through criticality due to failure is more practical.

1.2.3 Water Distribution Network Rehabilitation Issues

The aging and deterioration of water mains often leads to pipe corrosion, stress resistance capacity decline, pipe breakage, bursts, water loss, head loss, water quality degradation, more maintenance costs and other indirect loss. The immediate observed consequences of deterioration are often as follows:

1.   Pipe Breakage Rate or Frequency Increasing

This in turn leads to increasing operational and maintenance costs, increasing loss of water and social costs (e.g. water supply service interruption, disruption of traffic). Except for these, the cracks also provide a possibility of contaminant intrusion and increase health risk. Some other supplementary facilities (e.g., bigger tank or reservoir) might be required in a low reliability distribution system.

2. Decreased Hydraulic Capacity

Pipe failure results in increased energy consumption and more pressure imbalance in upstream and downstream pipes.

3. Decline of Water Quality in the Network

This may result in taste, odour and aesthetic problems in the water supply and even public health problems in extreme cases.

Pipe deterioration brings not only a heavy economic burden (repair and other costs), but also significant social (e.g. service interruptions, traffic delays, etc.) and environmental (e.g. lost water and energy) impacts. With the development of urbanization and population increase, a water distribution system becomes more important and urban water asset management becomes more complicated as well.

The aging and decline of WDS is one of the main problems facing water utilities around the world and not matter what action is taken, there is a certain inevitability due to environmental conditions, external damage, and soil and pipe movements. Due to the inevitable nature of the problem, utilities are required to perform routine and daily pipe maintenance work. Although investments in public infrastructure (e.g. WDS) are increasing, the rate of deterioration is much higher than the speed at which it is being resolved. Hence, utilities around the world are looking for innovative and proactive ways to deal with this issue in order to get ahead of the curve.

Water distribution system corresponds to the major proportion of the water supply system, which contains a lot of assets, and the distribution facilities in water supply systems will account for the largest cost item in future maintenance budgets. Data from 2001 in the Netherlands, on annual investments in the reconstruction and expansion of these systems, is at a level of approximately US$0.5 billion which accounts for 48% of the annual investment in the Dutch water supply works in that year (Trifunovic, 2006). The investment needs to tackle deteriorating water infrastructure are immense and this has been estimated to be over one trillion dollars over the next 20 years for water and wastewater utilities (Selvakumar and Matthews, 2017). Limited by budgets and technology, no water utility can keep all the water assets performing as well as a brand new system.

Owing to different severity and importance of each component in the system, not all aging and deteriorated pipes can be replaced or rehabilitated at the same time. Priority rehabilitation is an optimal problem and is essential for decision makers. The ideal strategy should be to make full use of the pipe's economical lifespan or use the minimum cost to obtain the maximum benefit during its whole life. Meanwhile, safety, reliability, water quality and economic efficiency will be considered. Keeping them in a good or acceptable condition with limited budgets and present technology is always the issue considered by water utilities. Planning for water main rehabilitation and renewal is imperative to meet adequate water supply objectives. Faced with huge amounts of rehabilitation expenses, decision makers need a pipe condition assessment tool to determine which pipes are to be renewed in what priority. The ability to understand and quantify pipe deterioration mechanisms is an essential part of the planning procedure.

A study has shown that a lack of prioritization and investment prohibits proactive pipe rehabilitation and this results in more frequent pipe breaks, increased leakage and energy costs (Roshani and Filion, 2014). However, water utilities are now beginning to realise that a more proactive approach to the management and rehabilitation of deteriorating assets is more cost-effective in the long-term than a passive one.

The passive rehabilitation strategy mainly reflects the lack of effective monitoring of the condition of the network and the lack of a response plan to its possible failure. Namely, the problem will not be solved until a problem arises. However, the best opportunity to solve the problem has been lost. This strategy is simple and easy, but the main defects are clear:

(1) Lack of forward-looking and systematic goals for future maintenance and rehabilitation of the distribution network. Then, the maintenance and rehabilitation work will always be conducted passively without precautions.
(2) Minor problems can be discovered and resolved only when they become big and serious problems.
(3) Costs of maintenance and rehabilitation increase, but system performances are difficult to maintain and upgrade.

The main characteristics of the proactive management strategy are that the possible failures will be predicted and corresponding plans and measures will be made before the failure occurs. With the network system as an example, the active management strategy should be like this: the key and important pipes are to be found as the main objects of rehabilitation, which is based on pipe structural integrity monitoring (or pipe structural condition assessment) and hydraulic performance significance evaluation of each pipe. A proactive rehabilitation strategy usually involves three stages: pipe deterioration assessment (e.g. pipe breakage rate estimation or pipe structure condition assessment); pipe criticality assessment; and optimization of rehabilitation decisions.

Due to capital limitation, infrastructure rehabilitation is usually not only an engineering and technical problem but also a management and social problem. From the view of engineering, the valid and efficient technology is currently lacking in monitoring, survey, maintenance, renewal and management. Meanwhile, other problems, such as absence of data, high cost and insufficient utilization of current database resources also exist. From the view of management and society, capital shortage and the low level of management limit the application of current technology. It is necessary that comprehensive methodologies are developed to assist planners and decision makers to find the most cost-effective rehabilitation strategy. Such a strategy should be based on the full extent consideration of a pipe's whole useful life while addressing the issues of safety, reliability, quality and economic efficiency. Optimization of design, operation and maintenance has always been, and will continue to be, the key challenge of any water supply company. Nowadays, accompanied with population explosion, the challenge is more underlined, particularly in some developing and newly industrialized countries. In the strategy of rehabilitation decision making, not only should the deterioration situation and tendency be fully understood, but the multiple stage, multiple aspects effect and consequence after rehabilitation should be judged correctly as well.

Generally, more investment brings more benefit. However, intelligent strategy is really needed because of budget limits and diversified objectives. Pipe rehabilitation can be regarded as the combat to the water distribution system's deterioration, subject to available budget and technology constraints. This will be intensively discussed in Chapter 5.

1.2.4 Asset Management of Water Distribution System

To deal with deterioration and aging of huge amounts of WDS asset, only efficient asset management (e.g., rehabilitation based on strategic planning) can systematically delay deterioration or partly recover asset's functions. In a general sense, asset management is the assessment of investment and management of them, so that their value is optimally enhanced while at the same time providing a benefit to the owner (Stephenson, 2005). Ofwat (the Water

Services Regulation Authority in England and Wales) provides another definition *(Ofwat, 2003)*:

"Asset management is a planning process that ensures that the owner gets the most value from each of the assets and has the financial resources to rehabilitate and replace them when necessary. Asset management also includes developing a plan to reduce costs while increasing the efficiency and the reliability of the assets. Successful asset management depends on knowing the system's assets and regularly communicating with management and customers about the system's future needs."

Strategic planning is usually embedded in asset management. Strategic planning allows decision makers to prepare in advance for unforeseen events. It involves using asset management data to understand and evaluate the status quo with respect to the physical condition of the assets and the institutional and financial capacity to deal with unforeseen events (Ofwat, 2003).

Good asset management strategy consists of both reactive and proactive strategies based on the condition and significance of pipe asset. A reactive strategy is a 'wait and see' approach whereas a proactive strategy aims to predict and anticipate future failures, and takes early, cost-effective actions to avoid such failures or to minimize their future consequence.

Asset management strategies for assets with different failure frequencies (or probability of failure) and different consequences of failure should be different (Burn *et al.* 2004, Moglia *et al.* 2006)). The "low probability and high consequence" pipe failures are usually managed using a proactive strategy and the "high probability and low consequence" pipe failures are usually mitigated using a reactive strategy. The "low probability and low consequence" pipe failures are usually managed using a reactive strategy which means leave it working until it fails. The "high probability and high consequence" pipe failures are usually to take action immediately. However, the boundary between low and high is unclear. Therefore, for most pipes with medium failure probability and medium consequence, reactive and proactive model based strategy is needed.

Generally, three themes are needed in asset management for a water distribution system.

1. Asset Data

Asset data can be categorized into static data and dynamic data. Static data (e.g., pipe material, diameter) do not change with time. While dynamic data usually refer to the data that change with time (e.g., broken pipe, rehabilitation). Data are the essential foundation for asset management. Lack of data and the inadequate detailing of available data is the most serious constraint affecting the application of generic asset management. Usually, there are some management problems in some water utilities. One is the data, especially dynamic data, are not so complete. For example, pipe failure and corresponding repair or renewal are not recorded in inventory. Another one is the inventory record form prevents the further utilization and analysis of the data. Old records usually are on paper. In order to utilize these data, there are several stages to transform the data on paper into digital form. The process is not only time consuming but also increases the possibility of make mistakes.

Over the past three decades, there has been great strides in relation geographic information systems (GIS) and global satellite positioning systems (GPS), and these technologies are of great benefit to water utilities as it allows them to more systematically and effectively map and record their assets.

A future vision is one where all relevant urban utilities will have access to accurate and up to date information on above-ground and underground assets in real-time, to enable them to take a more collaborative and coordinated approach to integrated asset management.

2.   Pipe Condition Assessment

An important component of an effective asset management strategy is the assessment of the condition of the assets. Data about the condition of the asset are often obtained through a few samples. As inspection of assets (particularly underground assets), is time consuming and costly, there is a need to develop a methodology that observes or even predicts the condition of pipes, based on numerous system descriptive parameters (indicators), which do not need direct inspection.

Existing models for predicting the conditions of buried pipes are based on statistical or physical/mechanical analysis methods. All the models are highly dependent on extensive data which need long-term collection. Research is required that identifies the major factors corresponding for pipe deterioration and the correlation between these factors and pipe failures.

3.   Decision Support System (DSS) Components for Asset Management

In order to improve water supply service levels, optimal strategies are necessary. Furthermore, a decision support system (DSS) based on optimal strategies should be developed that allows all possible alternatives to be considered, each with different costs and benefits. This DSS involves two main components, a criticality analysis that identifies the most critical components within the water network, and then a whole life costing (WLC) optimisation process to help identify the optimal replacement and rehabilitation strategies.

1.3 Research Contents and Technical Roadmap

In this section the research objectives of the thesis are presented and this followed by a description of the thesis structure and an illustration of the various components of the developed optimisation model.

1.3.1 Research Objectives

The overall objective of this thesis is to develop an efficient and effective whole-life cost optimisation model for the rehabilitation of water distribution systems. It aims to improve on previous models in that it attempts to combine critical features of the water distribution systems to inform and influence the optimisation search process in a way that improves convergence characteristics and overall computational effort.

The sub-objectives include the following:

(1) To develop a pipe failure (breakage) number prediction model that relates failure prediction to the important pipe characteristics/influence factors for homogeneous groups of pipes. This model plays a central role in the development of the objective functions of the whole-life cost optimisation model.

(2) To develop a pipe criticality assessment model that combines individual pipe condition assessment with a hydraulic significance assessment. The criticality assessment allows for the identification of the most important pipes in system (in terms of their condition and carrying capacity (hydraulic significance)), and for this information to be used to reduce the overall size of the optimization problem and inform the optimization search process.

(3) To develop an optimal rehabilitation decision model to help decision makers to understand the comprehensive performances in a network's whole life.

1.3.2 Technical Roadmap and Thesis Structure

Figure 1.4 illustrates the structure of the thesis.

Chapter 2 of this thesis presents detailed literature review of water distribution rehabilitation strategies, methods and models. It includes a review of water deterioration models and rehabilitation strategies based on the concept of whole life costing.

Chapter 3 of the thesis presents an innovative approach to the development of pipe failure (breakage) number prediction model. Major influencing factors that affect deterioration (i.e. pipe material, pipe age, pipe length, pipe diameter, freezing index and historical break record) are presented and discussed and an algorithm proposed to group pipes into homogenous groups. Using the influence factors and the pipe groups, a multiple non-linear regression approach is applied for the development of the pipe failure (breakage) number prediction model. The model is key to developing the objective functions in the whole-life cost optimisation model developed in Chapter 5.

Chapter 4 of the thesis presents a pipe criticality assessment model. This model combines a pipe condition assessment model and hydraulic significance model to establish pipe criticality. A modified TOPSIS method is applied to combine the pipe condition assessment with the significance assessment to generate pipe criticality indexes. Objective weighting assignment methods are proposed that avoid subjective judgment. Indicators are developed that provide a basis to choose a rehabilitation method - replacement or relining. The pipe criticality assessment model is used to reduce the dimensions of the whole-life cost optimization model of Chapter 5, in that it helps identify a subset of critical pipes for the optimization search to target and focus on.

Chapter 5 of the thesis presents the development of the whole-life cost rehabilitation optimisation model. Through the consideration of various potential objective functions, it develops two objectives (i.e. total burst number and modified resilience index) that are then used in the optimisation. Available budget considerations are taken as the direct cost constraints in the optimisation model. The decision variables are renewal actions for pipes including replacement, relining or no action. A genetic algorithm approach (a modified NSGA II), is used to solve the developed optimization model.

Chapter 6 of the thesis describes a case study application of the developed whole-life cost rehabilitation optimization model. The data for the model was obtained from a UK water company. Each component of the developed models (pipe breakage number prediction model, pipe criticality assessment model), are applied and used to develop the optimisation model which is then solved using a modified NSGA II. Through optimization model application, a set of Pareto solutions is presented and their dynamic performance with a renewal-deterioration cycle in the future are considered.

Chapter 7 of the thesis summarizes the work carried out and outlines future work.

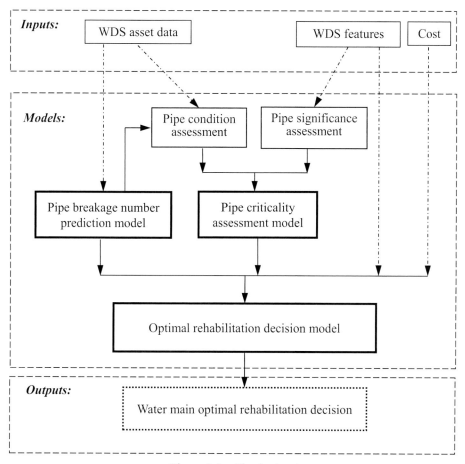

Figure 1.4   Thesis structure

Chapter 2   Water Distribution System Rehabilitation Strategy and Model

2.1 Introduction

The pipeline rehabilitation strategy is an effort to prevent deterioration and delay this natural trend systematically and with planning. Pipe breakage, burst, leakage and insufficient nodal pressure due to water main deterioration are usually the most important and immediate reason for rehabilitation, but the fundamental purpose is to improve the system's performance instead of some individual water main's structural integrity. A good rehabilitation decision strategy needs deep thinking and further understanding of the pipe deterioration process and the wide range of consequences of an action being taken or not. However, rehabilitation decision-making usually faces various difficulties, e.g. data deficiencies, great uncertainty, multiple objectives and the infinite rehabilitation-deterioration cycle.

There are numerous decision models to solve rehabilitation decision problems. The models concern at least one of the system's performances, i.e. economics, hydraulics, water quality, and reliability performance. An ideal model should account for all of these factors but the complexity and computation load make none of these existing models perfect. Different models with varying complexities concern different objectives and their combinations. The rehabilitation decision models in literatures can be classified into three categories: pipe deterioration model, pipe criticality assessment model and water main optimal rehabilitation decision model.

1.   Pipe Deterioration Model

Pipe deterioration process and its mechanism is an important foundation for rehabilitation decision making. The performance of pipe burst or breakage has been always the symbol of pipe aging deterioration. However, the occurrence of pipe burst can be regarded as the supposition of general deterioration tendency and some random destructive factors.

Accordingly, the general deterioration tendency is always covered by some apparent accidents. Pipe condition assessment involves models for failure prediction coupled with an understanding of processes that lead to failure, in order to predict future failures. The application of such models combined with field knowledge, historical records and inspection results, allows decision makers to make more intelligent, strategic and cost-effective decisions with respect to pipe rehabilitation and replacement (Liu and Kleiner 2014). Physically based models focus on the inherent mechanism and provide some convictive analysis but data deficiency is the bottleneck of the models. They are mainly used in some backbone or large size water mains. Statistical models are more widely used and developed because it is free from complicated mechanism explanations and rigorous data requirements. Data mining models are newly developed models. Whatever the model is, the results of the model calculations should be viewed as a statistical conclusion. Although it is possible to predict the pipe breakage trend and probable breakage numbers for pipes with some homogeneous characters, precise prediction for a specified pipe is difficult because of the great randomness.

2. Pipe Criticality Assessment Model

To improve an entire system's performances is the ultimate goal. Therefore, except for some poor structural condition pipes, pipes that have significant impact on hydraulic performance also should be identified and might be prioritized for rehabilitation. In addition, some pipes with insufficient water delivery capacity are to be replaced by larger pipes as well. The pipes that might be prioritized for rehabilitation are measured by these factors. The assessment of priorities can also be interpreted as an assessment of the criticality. The selected pipes with corresponding rehabilitation actions (e.g., replacement or relining) forms a set but their further impact on the entire system needs further analysis.

3. Optimal Rehabilitation Decision Model

Pipe breakage number prediction and criticality assessment models are the technical basis for developing further rehabilitation strategies. These strategies also include consideration of various costs and benefits. Moreover, the object is not the individual pipe, but the entire network system. Therefore, one of the characteristics of this study is to look at the problem from a systematic and global perspective, rather than performance improvement of an individual pipe or costs minimization only.

Optimal rehabilitation strategy is based on present and future pipe deterioration assessment. Water mains breakage number minimization will be one of the main optimization objectives because this is the direct motivation of rehabilitation and it has broad impact. Cost minimization is the main objective in most optimization models but it might not be the main objective if to pursue benefits or better performance with the limited budget becomes one of the main objectives in this study. Budget can be viewed as a constraint in optimization decision making if there is no immediate and explicit relationship between expenditure and performance improvement. Evolution algorithm is a powerful tool to solve optimization problems and it is applied in optimal rehabilitation decision making. The optimal rehabilitation with the consideration of whole life cost requires that the decision must be a multi-objective optimization based on present and future situations.

This chapter reviews the existing strategies and models of water distribution network rehabilitation decision, which includes water distribution system deterioration and pipe condition assessment models, pipe criticality assessment model, and optimal rehabilitation decision model. Through the models review, some drawbacks and challenges are identified and characteristics and improvements of new models are proposed.

2.2 Pipe Deterioration Models

Before a pipe rehabilitation decision is made, it is necessary to evaluate the deterioration degree and the condition of the pipes before systematical analysis. This usually needs to be done by modelling. For a large number of buried pipes, aging pipe material is the internal cause. External environment impact, such as mechanical (e.g. impact damage), physical (e.g. temperature and humidity changes) and chemical (e.g. corrosion damage is the external conditions) factors are the external causes of pipe deterioration. Because of different climate conditions, geographical environment, geological conditions and the physical and chemical environment, the mechanisms for pipe decline are very complex. It is also difficult to observe deterioration development directly. The research in this field is to evaluate the structural integrity of a large number of buried pipes, either directly or indirectly, by modelling. Because of the complexity of the mechanism and the diversified influence factors, coupled with the difficulty of direct observation, the method of modelling has been studied.

The pipe structural integrity is the key point of pipe condition. Namely, the structure of a pipe is perfect if there is no flaw (such as cracks, corrosion holes) on it. If a pipe is worn, its condition is evaluated according to the degree of damage. The collection and analysis of relevant data is the first and a main step to detect and monitor critical indicators to prevent or mitigate pipe failures.

2.2.1 Model Review

Comprehensive reviews on structural deterioration of water mains are from two aspects, respectively: statistical models and physically based models before that time (Kleiner and Rajani, 2001; Rajani and Kleiner, 2001). Later, Clair and Sinha (2012) carried out a state-of-the-technology literature and practice review on water pipe condition, deterioration and failure rate prediction models between the models found in literature and those currently used by utilities around the world. There are different techniques and methods for modelling pipe breakage based on identifying breakage patterns using statistical or data-mining (driven) techniques.

The general pipe deterioration model, or pipe condition assessment, or rehabilitation guides models usually propose a guide, such as an indicator addressing pipe structure condition or probable failure rate, for the identification of mains that require rehabilitation. For example, Shamir and Howard (1979) proposed an exponential function to describe pipe breakage rate growth. It treated water mains individually instead of systematically. However, these early models had some common drawbacks. For example, some important performance indicators, such as the hydraulic capacity, deterioration time, reliability, water quality and breakage of replaced pipes are not considered. Moreover, other rehabilitation approaches (e.g., relining) and energy cost were not always considered in the researches of the 1980s. The replaced pipes were usually assumed not to deteriorate.

The existing models to estimate water main deterioration can be classified into: (1) physically based models, (2) statistical models, and (3) data mining models. Below is a detailed discussion of these models.

1.   Physically Based Models

Physically based models usually focus on pipe wall corrosion process and mechanism (e.g. Chukhin et al., 2014), the residual stress resistance capacity and stresses from load applied to water mains. For example, electro-chemical corrosion is the main cause of exterior corrosion of cast and ductile pipes. This type of corrosion leads to the formation of corrosion pits that grow over time and ultimately lead to a pipe break. There have been several physical deterministic models developed to estimate the formation of corrosion pits and their impact on pipe strength (e.g., Kleiner and Rajani 2011, Kleiner and Rajani 2013).

Rajani and Kleiner (2001) classified physically based models into physical deterministic-and physical probabilistic-based models. The deterministic models take no account of the uncertainties in the deterioration and failure process, while probabilistic models provide insights into the contribution of each parameter to indicate the uncertainty of the result.

Seica and Packer (2004) developed a finite element model that used material properties obtained from experimentation, to estimate the remaining strength of the water pipes.

Clayton et al. (2010) studied the damage of clay shrinkage stress to the pipe. The calculated maximum tensile stress increase was found to be significant in terms of the residual strength of a corroded cast iron pipe.

Jesson et al. (2013) discovered that the strength of the cast iron pipe samples decreases with increasing depth of graphitisation through data analysed using the Weibull method.

Measuring strains in the pipe wall due to live loading can allow the estimation of circumferential flexural rigidity of pipes buried at different depths and in different configurations (Garcia and Moore, 2016).

Fluid transients are stress waves in the fluid that can propagate through pipelines and can collect information on the pipe condition during its travel. Covas and Ramos (2010) provide an in-depth analysis of the effectiveness of inverse transient analysis (ITA) for leak detection in WDS. They conclude that the application of ITA is limited as it requires an accurate description of the physical characteristics of the pipes, that the leak is of a 'reasonable size', and that the transient solver is accurate enough to describe the transient event. Hydraulic transients (e.g. water hammer waves) can be used to excite a pressurized pipe, yielding the frequency response diagram (FRD) of the system. Gong et al. (2013) used the FRD of a pipeline system for condition assessment and fault detection because it is closely related to the physical properties of the pipeline. Lee et al. (2015) applied a non-intrusive fault detection technology for real time condition assessment of pipelines. The results demonstrate that higher bandwidth signals provide more accurate fault detection at the expense of the detection range.

Some researches pay attention to the influence of pipe joint mode on the integrity of pipe network (e.g. Arsenio et al., 2013).

Generally, physically based models need intensive, detailed and large amount of on-site data and correct mechanism analysis. Some necessary data may be monitored, but much of the data required for physically based modelling is unavailable or very costly to acquire. Thus, physically based models may currently be justified only for major transmission water mains, where the cost of failure is significant. For other inferior parts, statistical models are more useful because they have flexible requirement for data. With the development of observation technology, this kind of model has a new development.

2.   Statistical Models

Statistical models are usually used to explain, quantify and predict pipe breakage, pipe structural failures probability and life expectancy. The division of pipes into groups with homogeneous properties (operational, environmental and pipe type) is often used, which requires efficient grouping schemes to be available. Statistical model usually predict the probabilities or frequencies of pipe failure by using asset and historical break data on the premise that future pipe failure follow the same rule as before (Scheidegger *et al,*. 2015). Some of which are foundation of later developed data mining models that have good prospects. Not only for the water supply network, but also for the drainage network, such models are applied (e.g.,Petit-Boix *et al.* 2016; Post *et al.* 2016).

Yamijiala *et al.* (2009) compare different statistical regression models (i.e., the time linear models, time exponential models, and Poisson generalized linear models (GLM)) proposed in the literature for estimating the reliability of pipes in a WDS on the basis of short time histories. The goals of these models are to estimate the likelihood of pipe breaks in the future and determine the parameters that most affect the likelihood of pipe breaks. The results show that the set of statistical models previously proposed for this problem do not provide good estimates with the test data set. However, logistic generalized linear models do provide good estimates of pipe reliability and can be useful for water utilities in planning pipe inspection and maintenance.

Statistical models attracted much attention from engineering practitioners and researchers because of its mathematical foundation and good adaptability to data. The statistically derived models are not critical with input data and can be applied with various levels of input data. It should be noted that statistical models also rely on an abundance of data but have more adaptability. Whatever the data, abundant or not, will not prevent the operation of the model but the accuracy and reliability of the model depends much on the data and model. These methods could help support reasonable and economical decision of rehabilitation/replacement in the present and future (Kim *et al.* 2012).

3.   Data Mining Models

Data mining involves searching for patterns in large data sets to enable the extraction of information from the data set and to develop the most important relationships within the data set. It can be considered as the analysis step of the "knowledge discovery" process from databases.

Statistical models can be regarded as a special type of data mining model, which uses statistical principles to process data. The application of statistical models requires a certain prior knowledge before testing the probability distribution and the statistical parameters. The data mining model does not require any prior knowledge, and can establish strong or weak links for a large number of seemingly non-directly related data.

The data mining models organize data and approach it differently so that useful information can be extracted. Usually, pipe material, pipe age, pipe diameter, pipe length, historical breakage record, or even historical winter temperature are available for most of water utilities. Although pipe deterioration influence factors are more than these, pipe deterioration information can still be extracted from these routine data.

Data mining techniques can be used for different purposes in management of a water distribution system. The commonly used Data Mining Techniques (DMT) in WDS analysis include artificial neural networks (ANNs), genetic algorithms (GAs), probabilistic and evidential reasoning, and fuzzy techniques. With respect to models that predict pipeline failure rates, artificial neural networks (ANNs) appear to perform better than statistical techniques. However, as ANNs is a black-box method, it's not helpful in establishing specific relationships between the variable concerned. (Bubtiena *et al.* 2012). GAs can be utilized for optimisation of system design, operational decisions, and maintenance plans. Fuzzy based techniques were used for pipe condition assessment (e.g., Yan and Vairavamoorthy, 2003) and failure risk assessment (e.g., Salman and Salem, 2012; Al-Zahrani *et al.,* 2016).

Since the deterioration process of the pipe has obvious randomness and a certain degree of no after effect, Markov process theories are used to describe such a process. A semi-Markov process is used to model the deterioration of a buried pipe (Kleiner, 2001). The life of the pipe is discretized into condition states, whereby the waiting time in each state is assumed to be random variables with known probability distributions. A Markov based approach is a decision support system to predict the future condition of a water distribution network (Sempewo and Kyokaali, 2016). Pipe condition has been based on a composite index that combines pipe age and break history.

Fuzzy set theory was applied in deterioration assessment as well. Yan and Vairavamoorthy (2003) and Vairavamoorthy *et al.* (2006) employed fuzzy composite programming (FCP) to make pipe condition assessment by interpreting 20 first level indicators that contribute to the deterioration of a pipe. Two companion papers were presented to describe an entire method of managing risk of large buried infrastructure assets (Kleiner *et al.* 2006a, Kleiner *et al.* 2006b). One uses a fuzzy rule-based, non-homogeneous, Markov process for modelling pipe deterioration. (Kleiner *et al.* 2006a). As this is a fuzzy-based model, it allows us to capture the imprecision and subjectivity that is inherent in pipe asset data. In addition, the model has the advantage of allowing the possibility of failures to occur during entire life of the asset. Another paper describes how the fuzzy condition rating of the asset is translated into a possibility of failure (Kleiner *et al.* 2006b). In these two papers, deterioration rate assessment only depends on pipe age and condition stage. Other influence factors are not involved. Kleiner *et al.* (2006 a) pointed out that pipe network deterioration assessment often face a series of problems, such as data scarcity, uncertainties and subjectivity of data, imprecision in cause-effect knowledge, and observation and criteria are expressed in vague (linguistic) terms. Fuzzy set and fuzzy-based methods are useful in dealing with many of these problems.

The advantage of the fuzzy-based method is dealing with the non-numerical parameters. These non-numerical parameters can be converted into numerical data through experts' experience and intelligence. Thus, the model is capable of combining the numerical and non-numerical parameters. This is a cornerstone of the method and it also causes some controversy. If the experience from different experts is not unanimous, or experts lack real experience, such an approach does not work well and will affect the reliability of the results.

Bai et al. (2008) applied the Dempster–Shafer (D-S) theory for pipe condition assessment, as this theory can help combine the many factors that affect pipe condition (bodies of evidence) at different hierarchical levels to provide a reliable assessment of pipe deterioration. This method is suitable for small conflict of evidence. If there is a high conflict between evidences, the following drawbacks can be seen: (1) the result might be contrary to intuition; (2) it will lack robustness; and (3) it is sensitive to the distribution of basic reliability.

Berardi *et al.* (2008) used Evolutionary Polynomial Regression (EPR) to select formulae which limited the alternative expressions in two terms and their coefficients values in some specified values need some prior knowledge. Although this method enlarges the formula type's search scope and becomes more intelligent than other models with a fixed formula type, it still needs an expert's knowledge to specify some candidate values as the coefficients' searching scope.

Rowe *et al.* (2010) developed a set of comprehensive scoring rules using advanced root-square-mean mathematical principles that can be used to process, screen and prioritize field inspection data so that it can be successfully used quickly for corrective action, in a cost-effective manner.

Opila and Attoh-Okine (2011) developed a method that translates pipe statistical failure models into a mean time to failure (MTF) model for individual pipes. The MTF is then used to score the structural condition of pipes (using economic discounting) that feeds a risk-based asset management models.

Pipe age data and the probability of lifetime can be combined to help decision makers to assess the future replacement needs of WDS. However, accurate lifetime estimation is still difficult due to the limited knowledge about deterioration process for different pipe material and varied condition. For this reason, Malm *et al.* (2012) concluded that historical data provide a reliable prediction.

Some missing reliable data is a barrier to establishing the most suitable model for a particular application. To overcome this, Scheidegger and Maurer (2012) generated synthetic data using a network condition simulator (NetCoS). Using the synthetic data, deterioration models were calibrated and their results compared with predefined scenarios. Based on such synthetic data, deterioration models are calibrated and their results compared with the predefined scenario. Although the model is for sewer networks, the method and principle are also applicable to WDS.

Multi-criteria Evaluation method is also used in pipe condition assessment because of diversified influencing factors and complicated relations. Sargaonkar *et al.* (2013) used a Fuzzy Multi-Criteria Evaluation approach for the development of a pipe condition assessment model. The model combined physical parameters (pipe age, material, diameter), operational parameters (number of breaks and bursts, leakage), and environmental parameters (bedding condition, overhead traffic, vulnerability of pipes to contaminant intrusion). Based on historical data and non-hydraulic factors of pipe burst, the traditional statistical method and improved analytical hierarchy process (IAHP) were adopted to conduct a qualitative and quantitative analysis between these factors and pipe burst in a district (Li *et al.*. 2015).

The application of pattern matching techniques and binary associative neural networks for novelty detection in the data of sensors monitor system (i.e. flows, pressures and water quality) are also applied (Mounce *et al.* 2014).

Data mining models attempt to break through the barriers of immediate data deficiencies or imprecise and vague. Data mining technique can extract some useful information through exploring available data that are usually incomplete. Data mining is used to find out the correlation among some factors, and even some logical relationships among them, from a quantity of data that are interrelated to each other but the relationships are not clearly defined. The implementation effect of such technology depends not only on the method itself, but also on the quality and quantity of the data. In a pipe deterioration model, by using data mining technology, it is possible to avoid the complicated mechanism, and to establish the quantitative relationship between the influence factors and the results of pipe deterioration directly. Generally, due to data deficiency and its impreciseness, the reliability of the information source has a great effect on the final conclusions.

4.   Modelling Development Trend

(1)   More Influence Factors Involving

Some ignored influence factors in the past are of concern in the deterioration model. For example, Fuchs-Hanusch *et al.* (2013) considered the effect of seasonal climatic variance in moderate climate regions on pipe failures, and found that soil moisture effects are only slight. This study showed that the amount of successive hot days (AHD) correlates well with failure frequencies in the dryer climate zones of Austria.

(2) Development of Inspection Techniques and Technologies

Liu and Kleiner (2013) reviewed a variety of inspection techniques and technologies for structural deterioration of water pipes. This included conventional non-destructive inspection and advanced techniques such as smart pipe, augmented reality, and intelligent robots.

(3) Integration of multiple methods

The combined model of the physical model, statistical model and data mining will be developed greatly under different data conditions.

Xu *et al.* (2011) applied three different approaches for pipe failure. The first involved a statistical model (Weibull distribution) and the other two were based on genetic programming (GP). The three models were applied to assess the failure criticality of pipe segments in the Beijing WDS and the two GP models were more efficient.

Mechanism model is combined with data mining model. For example, Sorge *et al.* (2013) developed an innovative method for determining maintenance intervention (both timing and place), where he combined the remaining-service-life prognosis for pipelines, with structural load factor verification, technical condition assessments, geo-referenced analyses and detailed costing.

Tee *et al.* (2014 a) developed an advanced Monte Carlo based simulation method called Subset Simulation (SS) for time-dependent reliability prediction. The study indicates that SS method is a robust way to predict the reliability of pipes, in particular for low failure probability/rare failure events. It also concludes that during the service life of a pipe, corrosion induced excessive deflection is the most critical failure event, whereas buckling is the least.

Kimutai *et al.* (2015), recommends that when assessing pipe condition and its impact on deterioration rate, it is better to use a combination of models rather than a single one. In their study, they applied three statistical models to analyse different covariates - the Weibull proportional hazard model (WPHM), the Cox proportional hazard model (Cox-PHM), and the Poisson model (PM).

Expert systems have been applied in data mining. For example, BBNs (Bayesian Belief Networks) is used for analysing drinking water distribution system data through the application of machine learning techniques to facilitate data-based distribution system monitoring and asset management (Francis *et al.* 2014; Kabir *et al.* 2015).

Some research integrated physical mechanism and statistical methods. For example, a stochastic model is applied to establish the depth of corrosion in cast iron pipes and a time-dependent method is applied for predicting the probability of serviceability failure. This allows the prediction of the time the pipe will be unserviceable and requiring repairs (Mahmoodian and Li, 2016).

Recently, hierarchical fuzzy expert system (HFES) technique was applied to develop a deterioration model and Fuzzy Monte Carlo Simulation (FMCS) was used to model the probability of failure and developing the risk index distribution for each type of asset (Marzouk and Osama, 2017).

2.2.2 Model Drawbacks and Challenges

The difficulty of pipe deterioration models application lies in the fact that a large number of buried pipes cannot be fully and directly observed. Although there are some non-excavation detection methods to investigate a pipe, on which some pits or cracks can be found, these methods are time-consuming and expensive. They cannot be fully applied for a huge number of network assets, except for some significant water mains, such as trunk mains. For a large number of general pipes, the pipe condition can only be evaluated indirectly through other evidence.

1. Drawbacks

(1) Lack of Comprehensive Understanding of Pipe Deterioration Mechanism

A perfect physically based model would explicitly encompass all the inter-relations between the factors affecting pipe breakage. But this often makes the model complicated and difficult to understand. The physical-based models, although scientifically sound, are limited in their application due to limitations in existing knowledge. These models aim to predict pipe failure, by combining pipe condition with an estimation of stresses on the pipe (through environmental and operational loads). The limitation, is in our understanding of the physical mechanisms that lead to pipe failure as these are complex and not currently fully understood.

While data mining and mathematical statistic methods can be used to describe the mathematical relationships between the data, there is still the risk of making mistakes in data mining and statistical results due to a lack of understanding of the mechanism analysis.

(2) Incomplete Consideration for Influencing Factors

The deterioration process in pipes is complex and involves many factors. Although physically based models capture many of these complexities, our limited understanding of the deterioration processes and the lack of data pipe condition data, make the application of physically based models limited. This problem is further compounded by the lack of appreciation of water utilities in the need to invest in more systematic and regular data collection activities.

(3) Depending Heavily on the Data Quantity and Quality

The data associated with the WDS update are not only spatially distributed but also dynamic. Therefore, it is difficult to obtain all the data needed effectively. The lack of total data or the lack of some key data will directly affect the effectiveness of the model.

(4) Influence of Experts' Experience and Preferences

Expert experience and preferences can largely determine the judgement and outcome in a class of data mining methods, such as fuzzy sets theory. If different experts have different views and understandings of the same facts, the resulting analysis may be quite different. Moreover, experts' bias and misconceptions also have a direct impact on the outcome. Therefore, the application of such methods must be based on consensus among different experts. If there are obvious conflicts of opinions, this approach does not work.

2.   Challenges

(1) Lack of Data

Data availability is still the major barrier for modelling. No pipe condition assessment model can provide a crisp and accurate conclusion due to data's spatial and temporal variety and large amounts of unknown/unavailable data. Although more data has become increasingly available from inventory databases (particularly the physical indicators), some data of other indicators are still difficult to obtain in practice (e.g. the external protection, workmanship, soil condition indicators) due to incomplete data records. Sometimes the needed data are diversified and data request is outreach most of water utilities' database.

The difficulties in data collection are as follows: (1) Some data are unavailable due to cost or spatial and temporal variability; (2) The observation horizon is not long enough; (3) The percentage of annual pipe failure to the total pipe number is small; (4) The annual deterioration change is not very obvious compared to the long deterioration period. For these reasons, some influence indicators have to be neglected or simplified in a deterioration model.

(2) Complicated Mechanisms and Influence Factors

The mechanism of pipe deterioration involves the corrosion of material, the change of stress condition of pipe material and the change of physical environment around pipeline etc. Because different interpretations can be given by different disciplines, this is a typical inter-disciplinary problem. Moreover, the different materials, installation conditions and specific environment of different pipes in practical engineering are very different, and the main mechanism leading to the deterioration is different. Therefore, it is difficult to make a unified and comprehensive mechanism analysis.

2.2.3 Characters of the New Model

(1) The foundation of model application must be long-term and planned observation and data accumulation. If the observation period is too short, the development of pipe deterioration is not enough, or only some incomplete data have been accumulated for some deterioration stages. Therefore, during the short observation period, the general pipe deterioration tendency is not yet clear. Long-term, planned monitoring is the only solution for the lack of data. Otherwise, the results of "rubbish-in, rubbish-out" will occur. The modelling technology itself cannot solve such a problem fundamentally.

(2) For a particular pipe network system, some key mechanisms and key influencing factors should be grasped instead of all the complex mechanisms and diversified influencing factors being treated equally. Although there are various factors and complex mechanisms resulting in deterioration, some dominant factors and mechanisms are still need to be identified. These key factors are useful to improve the efficiency of modelling.

(3) The expression of pipe condition assessment should be a crisp value because the results are further used for pipe criticality assessment and optimization decision. Otherwise, not only is the calculation workload is increased, but the final decision will be filled with a lot of uncertainty, by which the selection difficulty of decision makers will increase.

2.3 Pipe Criticality Assessment Model

Pipe criticality assessment model can be known as a prioritisation model, which is a problem oriented method. Models attempt to prioritise the water mains requiring rehabilitation and replacement. Namely, they rank the urgency of the mains to be renewed. The criticality of a pipe determines its rank or urgency of renewal. That is, where the pipes having a critical impact on the overall performance of the pipe network, the renewal and maintenance of the pipeline is given priority.

This kind of model mainly focuses on two key factors, which are pipe condition and pipe significance. The former is decided on by the integrity of the physical structure of the pipe, and the latter implies the influence of a pipe in a system if it works or not. These two are independent of each other.

2.3.1 Model Review

These models are usually based on different scenarios of the combination effects of basic factors in physical, hydraulic and experimental categories. Therefore, when prioritizing rehabilitation schemes, a combination of hydraulic analysis and the breakage models are usually applied (Tabesh and Saber, 2012).

These models can be further classified into two categories:

1. Prioritising Component Rehabilitation Models

These models often focus on a single performance measure (e.g. the pipe condition assessment, pipe failure rate, pipe failure probability, pipe service life and pipe vulnerability). Most pipe condition assessments or pipe deterioration models can be categorised into these because the outputs of this model is usually the index of which pipe should be given priority for replacement.

Pipe deterioration degree is indeed getting more attention in some priority models but this view is biased. Lack of systematic thinking is another prominent weakness of such models. The other is that performance improvement becomes a by-production of this type of model.

2. Systematic Pipe Criticality Assessment Models

The urgency of pipe replacement is based on criticality, which is the combination of the condition of the main and its significance in the network. Pipe deterioration degree is consistent with failure probability. Pipe significance has similar implications with failure consequences. In some research, vulnerability also refers to failure probability. Similarly, risk and criticality has the same implication.

Water distribution network components with high significance and poor condition need urgent work. Those with low significance and poor condition need a rehabilitation program but it is not so urgent. Those components with high significance and good condition need monitoring. Only those with low significance and good condition do not need any action. These decision rules are conceptual ideas and the real challenge is the measurement and boundary between the high and low significance, good and poor pipe condition.

Piratla and Ariaratnam (2011) proposed a relative criticality index (RCI) to quantify the relative criticality of pipelines by summing up the effects of reliability, cost of break repairs, and energy required to repair breaks in pipelines.

Rogers (2011) proposed a performance-based approach to estimate the present and future condition of pipes. This approach combines readily available data (pipe inventory and break record data) with site-specific parameters that avoid the need for high levels of technical expertise common to physical-based models.

Karamouz et al. (2012) summarized a set of factors being used to predict the vulnerability of each pipe and then classified based on severity of probable failure. The different levels of water distribution networks' vulnerability will help to set the priority of rehabilitation and maintenance activities in different parts of the system. For developing vulnerability zones, an evaluation about the vulnerability of system components is needed.

Salman and Salem (2012) present an approach where risk of failure is obtained by combining consequence-of-failure with probability-of-failure, using simple multiplication, risk matrices, and fuzzy inference systems.

In some research, rehabilitation plans are more focused on subareas than on each pipe due to unpredictable variables (e.g. cut off time of water supply, number of traffic control times, and location of valves) (Yoo et al., 2012). Meanwhile, the implication of significance is changed as well, focusing on the connectivity of the network, a methodology based on spectral measurements of graph theory to establish the relative importance of areas in water supply networks   (Gutiérrez-Pérez *et al.* 2013).

A fuzzy-based decision support system (DSS) is developed to identify vulnerabilities that may cause system failure within a WDS. The failures include those related to structural integrity, water quality and hydraulic capacity (Al-Zahrani *et al.*, 2016). The model uses an aggregate vulnerability index to represent the likelihood of system failure.

Some models focus on the priority of water leakage monitoring in a single pipe or a WDS area. In the research of Lin *et al.* (2015), a cluster identification method (CIM) is proposed to establish a priority for leakage detection and to assess whether spatial clusters of high failure-prone areas exist.

The risk of a burst pipe can also be considered as an index for pipe priority, in which probabilities and consequences are regarded as the two pillars of risk. Choi and Koo (2015) proposed a water supply risk (WSR) assessment model, which was developed for determining the pipe burst probability, the impact of pipe burst, and the WSR calculated as the product of these two values. The pipe burst probability is a management indicator for the water provider, and the impact of pipe burst is a management indicator for the water consumer.

The research of Fox *et al.* (2016) evaluates the interdependence of leak hydraulics, structural dynamics and soil hydraulics, particularly considering the significance of the soil conditions external to longitudinal slits in viscoelastic pipe.

Del Giudice *et al.* (2016) used a small sample of failure data and a statistical approach for the prioritization of preventive maintenance strategies. The method develops correlations between the most important factors affecting the vulnerability of the network and this is used to develop a vulnerability map of the network. Although the study is for the sewer system, the principle still applies to the water distribution system.

2.3.2 Model Drawbacks and Challenges

1. Diversified Criteria for Criticality

There is no uniform quantitative criteria and methods for pipe condition assessment and hydraulic significance assessment, which are the two pillars of criticality assessment. Some models consider only one of the major factors (e.g. pipe conditions) as a criterion for criticality. This also leads to the diversity of criticality assessment content and criteria.

2. Uncertainty in Conclusion

Since the two pillars of the criticality assessment, which are pipe condition assessment and hydraulic significance assessment, are subject to random factors and lack of data, the conclusions of criticality with these methods are uncertain. From a technical point of view, such an uncertain priority assessment conclusion increases the calculation load of the subsequent optimization decision, although such an uncertainty expression is scientific and rigorous. Moreover, it is usually difficult for decision-makers to understand the criticality evaluation conclusions with certainty.

3. Lack of Comprehensive Understanding of Loss and Cost

The models do not consider the various losses caused by failures, and rarely consider the costs required for maintenance and renewal either. There is a lack of comprehensive consideration about the possible consequences of the renewal decision and the performance-cost ratio (or cost-benefit ratio) of the decision. In addition, the assessment is based on historical data and current situation, while it lacks foresight for future development.

2.3.3 Characters of the New Model

(1) Although there are a variety of criteria, the new model boils down to two basic aspects of the pipe, structural condition and the hydraulic significance.

(2) The expression of the evaluation results is presented in a certain manner in the new model so that the results of pipe criticality assessment can be used as a basic indicator of the subsequent optimization model screening for the critical pipe section. Although there is a great deal of uncertainty in the pipe criticality assessment, it is treated as a deterministic index so as to simplify the overall index system and decision process.

(3) The criticality assessment model is mainly a static and single view model. It can quantify the degree of priority of pipe renewal based on the pipe structural condition and the pipe significance. Moreover, it can simplify the process of renewal decision-making without considering the costs and consequences. Meanwhile, it can be a bridge from the pipe condition assessment to optimize decision. Among a large number of pipes, the model makes a preliminary selection of pipes which may become the renewal objects, so as to narrow the scope of optimization and reduce the workload of optimization calculation. A comprehensive understanding of the loss and cost will be further addressed in subsequent optimization rehabilitation models based on whole life costing.

2.4 Water Main Optimal Rehabilitation Decision Model

The optimization design of WDS has been a hot topic of WDS research for a long time. Initially, the research focused on a single objective optimization of cost minimization in WDS design. Currently, the optimization involves the combined design and rehabilitation process of multi-objective optimization with the whole life of WDS. The numbers of optimized objectives and the optimization algorithms have been greatly developed. In most researches, economic objective has been the only or one of the main objectives. The immediate cause is that it is the primary business motivation for water utilities. Meanwhile, it has clear definition and measurement. The algorithm of single-objective optimization is relatively simple and needs less computation load. The objective selection mainly depends on the major purpose of decision maker and computation capability. Most of the research on the WDS optimization focuses on the design, and some of them are about optimal operation. However, research on the optimal rehabilitation decision is relatively few. The reason for this may be that the renewal problem of WDS becomes prominent with the asset increase and deterioration. The research for such a problem is not enough because it appears relatively late.

The pipe criticality assessment and pipe optimization renewal decision are to solve the same problem: which pipes should be renewed and replaced urgently? The difference is that the pipe criticality assessment does not consider the costs and other consequence, such as performance improvement of the system. In the optimization decision of pipe network renewal, both issue of costs and consequences are considered, and the results of quantitative calculation are given.

2.4.1 Model Review

Optimization models are the further development of the prioritising models. For most existing water distribution systems, optimal scheduling of the maintenance and improvement for a long time is necessary.

The multi-objective optimization of a WDS takes the minimization of cost as one of its major objectives, while the benefits (e.g. the improvement of a performance or reliability) are considered as opposing objectives. This characteristic has not changed much since earlier studies. However, the meaning of benefit has different understandings in different studies. For example, Halhal *et al.* (1997) minimised rehabilitation costs while maximising the benefits of rehabilitation. The benefits include, hydraulic and water quality performance, financial savings, and increased reliability of the system. It can be argued, that except for the benefits associated with reduced bursts, the other benefits are subjective. Engelhardt (1999) also proposed a multi-objective function where he minimised operating cost while maximising reliability (expressed in terms of customer interruptions).

The development of multi-objective optimization research on WDS is mainly reflected in the diversification of optimization objectives and the improvement of optimization algorithms. Moreover, the model also is more deep thinking in dealing with uncertainties.

Osman *et al.* (2012) argues that many of these pipe condition assessment technologies are expensive and not reliable. They present an approach that combines cost of generating data using condition assessment technology with the value of the data generated. The incorporate this thinking into an optimisation model, that considers both direct and indirect costs of infrastructure failure.

Nazif *et al.* (2013) proposed a method that combines the condition of the pipe with its significance to the network. They developed two indexes: Physical Vulnerability Index (PVI) that evaluates the physical status of pipes; System Physical Performance Index (SPVI) that articulates the distance of the pipe from a reservoir and average pressure of pipe. However, the costs are not emphasised in optimization and the rehabilitation purpose mainly focuses on water main physical conditions, instead of some service performances, such as water pressure.

The study of Siew *et al.* (2014) presents a whole-life design and rehabilitation approach that involves multi-objective optimisation. The model considers both structural integrity and hydraulic capacity of the pipes. In terms of costs, it includes construction, rehabilitation and upgrading costs with pipe failure costs. The optimisation method applied is an evolutionary approach, where the fitness function is a trade-off between its lifetime costs and network hydraulic properties.

Tee *et al.* (2014b) estimated the reliability of non-pressure pipes experiencing externally applied loading and material corrosion. Estimation was provided on the expected time of corrosion induced deflection, buckling, wall thrust and bending stress and this was followed by recommendations on maintenance intervention including timing, and renewal solution (generated using a GA based whole life cycle cost optimisation).

To deal with the uncertainty in the development, the Markov decision process (MDP) based methodology to minimize the cost is proposed (Kim *et al.,* 2015).

There has been great interest in surrogate indicators for the hydraulic reliability and/or redundancy of water distribution systems. Tanyimboh *et al.* (2016), considered several surrogate indicators, including statistical flow entropy, resilience index, network resilience and surplus power factor. Their study showed that using statistical flow entropy, the reliability of the WDS can be estimated well with limited computational effort. They found that the other surrogate measures considered were often inconsistent.

Some studies have adopted the method of staged optimization. For example, the multi-objective optimization model is divided into three sequential stages, and the Pareto Front (PF) is gradually identified (Rahmani, *et al.* 2016). The first two stages involves analysing a skeletonized WDS, using a two-objective optimization model. The PF is improved from stage 1 to stage 2. The third stage adopts a three-objective optimization model to the full network.

Muhammed *et al.* (2017) present a new optimal rehabilitation methodology for WDS based on the graph theory clustering concept. The methodology starts with partitioning the WDS based on its connectivity properties into a number of clusters (small subsystems).

Similar to multi-objectives optimization, multi-criteria decision analysis is employed to make strategic rehabilitation planning of piped water networks. For example, in the research of Scholten *et al.* (2014), three fundamental objectives (low costs, high reliability, and high intergenerational equity) are assessed. The criterion of intergenerational equity reflects a long term view. Eighteen strategic rehabilitation alternatives under future uncertainty are evaluated. The commonly used reactive replacement is not recommended unless cost is the only relevant objective.

Both, performance in general and cost of rehabilitating the system, play a role in the rehabilitation programme. Such an approach allows for the trade-off between system performance and cost of rehabilitation. The multi-objective optimization techniques require large amounts of trial calculations to search for the near-global optimal solutions. Due to the diversity of performance requirements and costs, there is no perfect model comprising all of the performance requirements and costs. Many of the models currently only consider one or two performance objects because of the complexity and computation load. In most of cases, simplification is utilized when there is more than one objective. Multi-objective optimization approaches can formulate whole life costing models effectively and provide optimal trade-off between economic, hydraulic, reliability and water quality performance criteria.

The further development of modelling has to be concerned on how to deal with the complicated objectives and the uncertainty in the long term. In order to focus on the key factors and to deal with the conflicts between computation load in searching for the near-global optimal solutions and the multi- objectives optimization in models, the objective number have to be reduced if some comprehensive indicators are proposed. Although such a comprehensive performance indicator cannot be perfect or comprise all the necessary performance indicators, it should be representative, easily understood and calculated. Furthermore, a decision's long term impacts should be considered. The uncertainty in future development, which is often ignored in most existing researches, results in more complexity in decision making.

Rehabilitation is a different from newly designed and construction for a water distribution network. Because the performance requirement after rehabilitation is diversification, a long term view is needed, and cost saving is not the only object. Because budget limitation is usually a constraint in rehabilitation practice, maximum cost-effective, instead of maximum performance or minimum economic cost alone, is the main objective in decision making.

2.4.2 Model Drawbacks and Challenges

The motivation of the pipe rehabilitation is diversified and very complicated, such as pipe aging, pipe deterioration, increased water leakage, insufficient water supply capacity, and some water quality problems. Moreover, the rehabilitation strategy is not only to solve current problems, but also to provide opportunity and flexibility for future development. The various causes and forward looking make the rehabilitation strategies complicated. Generally, the drawbacks and challenges in rehabilitation strategies are summarized as follows:

1. Uncertainty in Decision Premise

Uncertainty widely exists in decision process. For instance, pipe deterioration and water demand cannot be accurately predicted. A reliable long-term system requires high reliability when designed. This principle is also suitable for rehabilitation decision. Thus, assessment of the network condition during the operational period can be an effective way to increase the network's efficiency (Seifollahi-Aghmiuni *et al.* 2013).

Uncertainty makes the decision precondition and consequences become more uncertain and causes more difficulty in decision making. Moreover, the optimization result lacks sufficient realistic meaning since the premise of optimization is uncertain and variable. It is difficult to make accurate judgement for the future at the current stage because of the randomness of a variety of incomplete information. The absolute optimization decision is only an ideal state. In practice, the goal should be to find one or more "no-regrets" decisions.

2. Diversified Motivations

The fundamental driving force of pipe rehabilitation is to improve system's general performance instead of to replace some poor condition pipes. If this fundamental driving force is ignored, the rehabilitation decision might be only suitable for some specific purpose. What a decision maker needs is a systematic view, which includes multiple performance improvement and high cost efficiency.

3. Multiple Stage Decision Process

The dynamic process of pipe deterioration and rehabilitation actions are often simplified. In some models, only pipe deterioration without rehabilitation is considered. In some other models, rehabilitation action is taken into account only once, instead of continuing work over a long time. Deterioration is a continuous process for all of the network components but deterioration in the future is not always considered, especially that of new pipes after replacement. One of the reasons might be the deterioration process or pipe failure is so complicated that no explicit function can describe or predict perfectly. Another reason might be that some functions involve too much uncertainty or some data is unavailable to making the application infeasible or very complicated. Combat between pipe rehabilitation and pipe deterioration has been existing since pipes have been laid down. Rehabilitation strategies and actions are taken every year and the system's performance is also improved gradually by such a driving force. Pipe deterioration, a continuous process, occurs all the time for each pipe. Nevertheless, rehabilitation actions are only taken with some critical pipes. Therefore, the deterioration-rehabilitation process is a chain scenarios of multiple stages, which causes a huge computation load. Particularly, the scenarios are very complicated if enumeration is applied to simulate the uncertainty.

4.   Diversified Costs

In terms of cost, only direct costs are considered, whereas indirect and social costs are seldom taken into account in a reasonable way in most of models. Because direct cost is easy to measure and relate to each party's economic benefits, more importance is attached to it. In contrast, indirect and social costs are difficult to be understood, quantified and accepted. For example, leakage results in pipe bed deterioration earlier in most of cases. Nevertheless, the leakage rate and its impact on pipe bed are still difficult to quantify. In addition, the direct costs are not always considered perfectly because of model simplification or some data being unavailable.

5.   Complexity of Algorithm

The multi-objective optimization algorithm, taking genetic algorithm as an example, has been improved in many ways in order to adapt to the problems with different characteristics. Meanwhile, the hydraulic calculation of pipe network is also improving. When the two are intertwined, both computation load and computation complexity increase significantly.

The existence of these challenges illustrates the need for further research on this issue.

2.4.3 Characters of the New Model

The most striking feature of the new optimal rehabilitation decision model is Whole Life Costing (WLC). This concept implies a multi-objective and multi-stage decision process to maintain a system to serve at a required level.

The concept of whole life costing was widely applied in civil engineering projects in the last decade. The term of whole life costing (WLC) was originally applied in building and structural engineering. The background to this term is the recognition that initial capital costs are only a small portion of the overall costs incurred during the lifetime of an infrastructure. The UK government advocates the adoption of WLC approaches to ensure that all costs are considered, and a way to optimising investments in cost-effective solutions.

The characters of whole life costing are as follows:

1.   Global View of the Cost

The costs include the initial capital or purchase costs required to establish the facility as well as operational, maintenance, rehabilitation, repair costs and the decommission costs. Moreover, it not only considers the direct costs but also the associated indirect costs as well. If the concept of cost is further extended, including indirect cost and damage cost due to failure, it is similar to the general costs of failure/deterioration.

2. Long Term View

Life time is a vague term without a definite time span. It is used to indicate either economic life, operating life, design life or useful life, whichever happens in accord with the purpose and use of the facility being considered (Skipworth *et al.,* 2002). For a water distribution system, the system's life may be as long as the city's life, but the system's components' lives are not so long. Not only the current impacts but also future impacts are to be considered.

3. Systematic View

Rehabilitation focuses on the overall network's performance improvement instead of one or a few pipes' structure or local performance improvement. Furthermore, long-term and far-reaching impact of the actions must be taken into account as well. In a macro view, the general rehabilitation objective is to improve or keep the system's performances (e.g. pressure and water quality request) being acceptable at relatively low cost. In a micro view, the objective is to combat with pipe structural, hydraulic and water quality deterioration so that distribution system's deterioration process can be slowed down. Budget limitation, service standards and hydraulic principles are major requests and constraints in decision making.

It is clear that the concept of WLC in both space and time dimension are greater than that in traditional views. The concept of whole life costing is widely applied in engineering for these reasons:

(1) Short term view may earn more economy benefits, but will damage long run benefits;
(2) Making trade-off is needed between short-term and long –term benefits;
(3) Different component have different life spans.   Hence, making a strategic plan and setting aside funding can solve the problem (Ofwat 2003); and
(4) Minimize social and indirect costs.

According to the characters of whole life costing, the main improvements of a new model can be summarized as follows:

1. Objectives

Pipe failure number reduction is one of the main objectives. Moreover, the capability of dealing with some accidents or unpredicted water demand growing is taken into account as well. The usual objective, monetary cost minimization, is converted into a constraint. In some cases, monetary cost minimization is not the absolute principle. To obtain the best performance from expenditure is the main purpose. Through the analysis of costs and benefits, some immeasurable parameters are converted into quantified surrogates.

2. Process

Both pipe deterioration and rehabilitation are simulated as an integrated and continuous process in the new model. These conflicted forces always exist in the whole life of WDS. Therefore, the simulated objects of the new model are the recurring process of "deterioration-rehabilitation". The system's performance is determined by the combat results of the two forces.

3. Optimization Algorithm

A group of near optimal decisions can be made based on an improved multi-objective optimization. Then, with the general deterioration tendency and possible consequence in prediction, some rehabilitation suggestions can be made. Such a repeated chain serial of deterioration-rehabilitation scenarios will be the background for optimization decision process.

2.5 Summary

The entire water distribution system rehabilitation strategy and model is composed of three parts: pipe deterioration assessment, pipe criticality assessment, and optimal rehabilitation decision making. The pipe deterioration degree is one of the most concerned issues for water distribution network renewal. Meanwhile, the predicted numbers of breakage for different diameter pipes are also needed to assess the costs in rehabilitation decision model. Pipe condition and pipe significance are two foundations of criticality assessment, which is the indicator of renewal priority. Renewal consequences are quantified by various costs and benefits in optimal decision making. Therefore, there are three interrelated models. A brief description of each model component and their relationship is given below.

1. Pipe Deterioration Model

Water distribution system deterioration models can be subdivided into physically based models, statistical models and data mining models. Physically based models usually need intensive observation data and experimental data in the local environment, which are mainly applied in small quantities and important trunk mains. Statistical models require extensive data and are applied more widely, and they usually need a lot of data to obtain higher reliability results. Data mining models, which integrate artificial analysis sometimes, may excavate more information according to the available data than the statistical models in some cases. These models partly compensate for the over dependence of the previous two models on the observation data.

The main drawbacks are addressed, which are lack of comprehensive understanding of pipe deterioration mechanism, incomplete consideration for influencing factors, depending heavily on the data quantity and quality and influence of experts' experience and preferences. The main challenges come from lack of data, complicated mechanism and influence factors.

Some key mechanisms and key influencing factors, instead of general ones, will be paid more attention to in the new model. The expression of pipe condition assessment will be a crisp value so that the results are further used for criticality assessment and optimization decision.

2. Pipe Criticality Assessment Model

Pipe criticality assessment model can be known as prioritisation model. The models can be subdivided into prioritising component rehabilitation models and systematic pipe criticality assessment models according to the complexity of rehabilitation strategic thinking.

The main drawbacks and challenges are addressed, which are diversified criteria for criticality, uncertainty in conclusion, and lack of comprehensive understanding of loss and cost.

The new pipe criticality assessment model can simplify the process of renewal decision-making without considering the costs and consequences. Meanwhile, it can be a bridge from the pipe condition assessment to optimize decision. The model makes a preliminary selection of pipes which may has potential to become the renewal objects, so as to narrow the scope of optimization and reduce the workload of optimization calculation.

3. Optimal rehabilitation decision model

The main drawbacks and challenges of optimal rehabilitation decision models are addressed, which are uncertainty in decision premise, diversified motivations, multiple stage decision process, diversified costs, and complexity of algorithm.

Optimal rehabilitation decision model is the specific application of the water distribution network renewal strategy. The most striking feature of the new optimal rehabilitation decision model is Whole Life Costing (WLC), which is a comprehensive and proactive strategy. The characters of whole life costing are global view of the cost, long term view, and systematic view. The main improvements of new model lie in objectives, process and optimization algorithm.

Chapter 3   Pipe Breakage Number Prediction Model

3.1 Introduction

Water distribution systems are a major component of a water utility's asset and may constitute over half of the overall cost of a water supply system. Pipe failures within the distribution system can have a serious impact to both people's daily life and to the wastage of limited, high quality water that has undergone extensive treatment. Hence it is important to maintain the condition and integrity of distribution systems.

Direct observation of a pipe's physical condition (e.g. cracks, crevices, internal incrustation), is not possible due to the fact that they are buried underground. Pipe deterioration assessment is therefore often derived indirectly through the analysis of pipe and environmental data.

In this study, methods are proposed to estimate deterioration at both the system level and individual pipe level and hence two independent models are proposed:

(1) Pipe breakage number prediction model, which groups pipes based on their common characteristics and then develops a model to predict the groups failure rate using regression analysis. This model is used in the development of the objective functions for the whole-life costs pipe rehabilitation optimisation model described in Chapter 5.

(2) Pipe condition assessment model, which attempts to estimate the condition of individual pipe's based on a suite of physical and environmental data. It combines this condition assessment with hydraulic significance of a pipe to generate a pipe significance index that estimates the criticality of individual pipes. This model is and helps reduced the size of the whole-life costs pipe rehabilitation optimisation model, by identifying key pipes for the optimisation to focus on during its search process. This model is described in Chapter 4.

In this chapter, a new pipe breakage number prediction model is presented, where pipes are grouped and then a relationship is established between the key pipe characteristics of the group (independent or predictor variables), with their respective breakage rates (dependent or criterion variable). The key independent variables in the analysis includes pipe material, pipe age, pipe length, pipe diameter freezing index etc. and the dependent variable is the historical break record.

In the preceding sections, descriptions and reviews will be provided on the main factors that influence pipe deterioration. This will be followed by a detail description on the development of the multiple regression pipe breakage number prediction model.

3.2 Pipe Deterioration Influence Factor

There are numerous factors that can influence a pipe deterioration process (Yan and Vairavamoorthy, 2003; Al-Barqawi and Zayed 2006), but there is no complete and comprehensive research which considers all the factors together and line them up solely based on their relative weights of importance (Zangenehmadar and Moselhi, 2016). Environmental characteristics and material properties can be key factors to pipe deterioration (Park *et al.*, 2016). Comprehensive research shows some indicators in Table 3.1. These basic deterioration indicators influence pipe deterioration most. The combined effects of these factors make water pipe deterioration rather complex. It is difficult to distinguish each factor's influence separately. However, the major influence of each different factor is described in the following.

Table 3.1   Water main deterioration indicator (after Yan (2006))

Physical indicators	Environmental indicators	Operational indicators
Material	Bedding condition	Frequency of supplies
Year of installation	Traffic load	Duration of water supplies
Diameter	Surface permeability	Number of valves
Length	External protection	Number of connections
Joint method	Soil condition	Leakage record
Internal protection	Groundwater table	Complaint frequency
Workmanship	Buried depth	Breakage history

3.2.1 Overview of the Main Influence Factors

1.   **Physical Indicators**

(1) Pipe Material

Pipe material is the crucial factor to pipe deterioration as it may affect the rate of deterioration with other indicators. It determines pipe corrosion resistance, impact strength and pressure resistance. The chemical property determines the corrosion resistance capability and water quality deterioration to some degree. The corrosion resistance implies the intrinsic ability of pipe material to resist degradation by corrosion. The physical property determines the stress (impact strength) and pressure resistance capability to the loads. The maximum pressure reflects the strength of pipe material. The impact strength represents the ability of a material to withstand impact without damage.

Cast iron pipes have been widely used in water distribution systems in history. However, nowadays, ductile iron is commonly applied. Rajani and Kleiner (2001) discovered that the deterioration of the exterior of cast and ductile pipes is mainly due to electro-chemical corrosion. The factors that accelerate this corrosion include, stray electrical currents, soil moisture content and other soli characteristics, chemical and microbiological content, electrical resistivity, etc. With respect to internal corrosion, the chemical properties of the water itself play an important role (e.g. pH, dissolved oxygen, free chlorine residual, alkalinity), as well as the temperature of the water.

Asbestos-cement and concrete pipes have been also widely used in history. The deterioration of asbestos-cement and concrete pipes is due to several chemical processes that relate the soil characteristics (e.g. organic or inorganic acids, alkalis or sulphates in the soil). For example, with reinforced and pre-stressed concrete, soils with low pH can affect the cement mortar to a point where corrosion of reinforcements take place, badly affecting the strength and integrity of the pipe. As the external mortar in the pipes further deteriorates, the steel wires get exposed and dwindle away, leading to eventual pipe failure.

PVC and other plastic pipe material is also popular. PVC pipes have been used commercially only in the last 45-50 years, and the material is a better corrosion protection. For these reasons, the long-term degradation mechanisms in PVC pipes are not well documented. It is generally believed that its deterioration is slower than that of metallic tubes. The deterioration mechanism may also be chemical and mechanical degradation, oxidation and decomposition plasticizers and solvents.

(2) Pipe Age (Year of Installation)

Pipe age is closely related to burst rate. This is in line with common knowledge. Moreover, the third period of "bathtub curve" also indicates the higher failure rate with pipe age. However, some studies have also suggested that age alone is a poor indicator of the necessity for pipe replacement or rehabilitation (Wang *et al.*, 2010).

Age, or more usefully date laid, however, does give an indication of the length of time that a pipe has been in operation, exposed to the surrounding environment and both internal and external loading. Because the erosion or attack from the environment accumulates with time, and pipe material also deteriorates with time, pipe age can be regarded as a surrogate of pipe structural health. Date laid may also be a surrogate for design and construction practices and the quality and strength of the material itself, although such factors would not be expected to result in predictable or smooth trends (Boxall *et al.*, 2007). Pipe age alone is not a good immediate indicator of pipe structure, but it is a good surrogate to show how much attack pipe suffer from the environment and how much self-deterioration it accumulates.

(3) Pipe Diameter

Many studies have a conclusion that small diameter mains (i.e. trunk mains less than 300 mm) suffer higher break rates than large diameter mains (e.g. Boxall *et al.*, 2007). The early modelling approaches of Shamir and Howard (1979) and Walski and Pelliccia (1982) have recognised inverse relationships between breakage rate and diameter.

The possible reason behind the diameter may be the pipe wall thickness and corrosion pits depth (Cooper *et al.*, 2000). Generally, pipes with a large diameter also have a thick pipe wall which can endure more stress and pressure if they have the same material and all of other environmental and operational conditions. Meanwhile, the same corrosion pits are a relatively lighter hazard to a thick pipe wall than to a thin one. Furthermore, small pipes lack sufficient bending strength and are susceptible to corrosion.

Another possible reason is that the installation of larger pipes is taken more seriously, due to their importance in the system and the effort required in their placement (Davies *et al.*, 2001). Ground movement is another possible reason. Larger pipes by their very nature are less susceptible to displacement relative to the ground, because they have larger cemented surface areas (Cooper *et al.*, 2000).

(4) Pipe Length

The length of a pipe may be expected to be related to failure number as a linear function. If pipes are laid in a homogenous surrounding, failure rate will not be impacted by pipe length and failure number will increase linearly with length. In practical pipes, this is not always the case due to the inhomogeneous surrounding. If a pipe is not long enough, its breakage occurrence is mainly affected by some random factors and the failure tendency with length is not very clear. If a pipe is long enough and other influence factors are quite similar, the breakage rate will not vary greatly. Therefore, the error of a short pipe's break number estimation is usually greater than that of a long pipe. Kleiner *et al.* (2007) also illustrated the obvious linear correlation that exists at the macro level between pipe length and the number of breaks, but that at the micro level this correlation is not clear and is dominated by noise due to the large and natural variation between pipes. In order to eliminate the randomness and to find the general pipe failure tendency, the pipe might be grouped by homogeneous features. In such a case, pipe failure numbers basically increase with pipe length and failure rates become stable.

(5) Pipe Joint Method

Different pipe materials often adopt different joint methods which can be classified into rigor and flexible joints. Some flexible joints have more scope to adapt for the movement of surrounding soil or the connected pipe sections. For example, the rubber gasket joint alleviates the shortcoming associated with leadite and rigid joints in terms of allowance for deflection, while a leadite joint is inferior to a lead joint (American Water Works Service Co., 2002). This is a non-numerical parameter with large spatial variability.

(6) Internal Protection

The pipes with an internal protection of lining and/or coating are not easily corrosive. Most modern metallic pipes have an internal lining to prevent internal corrosion from soft or aggressive water. However, older metallic pipe might be unlined or with a damaged protection layer, and thus susceptible to internal corrosion. Pipeline corrosion will lead to degradation (e.g. pitting), which may cause leakage or mechanical failure. This is a non-numerical parameter with large spatial variability.

(7) Workmanship

Workmanship deals with the human factor of the quality control of construction work. It is clear that poor workmanship may deteriorate the pipes and cause more risk, regardless of pipe age and other factors. However, there is no clear workmanship assessment in the construction record. The default is satisfying the standards and codes. This factor is seldom involved in this study but plays an important role in engineering practice.

2.   Environmental Indicators

(1) Bedding Condition

Bedding support is an important part of the pipeline installation and the pipe must be placed in a proper bed. Bedding type is determined by a number of factors, including pipe material, size, and surface load and working pressure. Ideally, a pipe should be supported uniformly over its entire length, although this may change over time due to disturbance. If a pipe lacks good support, there is a danger that it will act as a beam where it may experience shear stresses and bending moments. Its ability to resist such forces is a function of the pipe's material and geometrical proportions (Boxall *et al.,* 2007).

(2) Load

If external or internal load exceed a pipe's residual stress resistance capacity, pipe break will occur. Other than frost, traffic load is another major external load. Pipe failure rate increases with traffic load and traffic load is normally greater on principal roads. Generally, traffic load increases with traffic if the pipe's buried depth was not considered pipe failure occurs.

Under normal operation conditions, pipes withstand an acceptable water pressure. However, great pressure often created from surge events caused by, for example, water hammer from pumping switching or valve operation. Surge events have the potential to cause failure by exposing the vulnerable parts of the network, sometime repeatedly, to excessive pressures. This is the immediate and violent impact from internal pressure (Skipworth *et al.,* 2002).

Internal water pressure is probably the single most important factor controlling leakage. High water pressure provides not only more capacity to deal with unpredicted water demand or incident, but also results in more leakage. From a water main deterioration point of view, leakage will result if pipe bedding erodes and, in turn, pipes lose bedding support. A lack of support will result in a pipe acting as a beam, exposing it to sheer forces and bending moments. A pipe in such a situation is prone to break.

This is a numerical indicator but has great temporal and spatial variability. Its accurate record is not available in a water utility's inventory or asset database.

(3) Surface Permeability

Surface permeability is the degree to which water and moisture can percolate to the pipe. Surface salts carried to the pipes with the water an moisture, coupled with intermittent wetting and drying of the soil surrounding the pipe, will increase the risk of corrosion and ultimately the deterioration of the pipe. This is a non-numerical indicator and has great spatial variability. Its accurate record is not available for most water utilities.

(4) Groundwater Condition (Groundwater Table)

The pipe may be above or below the groundwater table or could intermittently be in both. Groundwater affects pipe deterioration in the following ways: (1) Chemicals in the water may be aggressive to the pipe material (2) Bedding support of the pipe may be negatively affected by the flow of water and result in poor support to the pipes. (3) Intermittent wetting and drying may make the bedding material unstable. This is a non-numerical indicator and has great temporal and spatial variability. Its accurate record is not available for most water utilities.

(5) Buried Depth

The depth at which the pipe is buried will influence its resistance to failure. Good pipeline installation, provides sufficient depth so that overhead traffic does not impact the structural integrity of the pipe. It has been reported that defect rates decrease as depth increases up to a certain depth after which the defect rate rises (Davies *et al.,* 2001). The reasons given for this is that the initial decrease is the result of the impact of overhead traffic, but as we go deeper, there are negative effects associated with the backfill, earth pressure and soil moisture.

(6) Temperature

Air temperature often leads to changes of soil temperature, and in turn results in soil expansion and contraction. Although this factor is not included in Table 3.1, its influence should not be ignored, especially in cold regions. In observation, frost often produces more pipe bursts.

It is discovered that seasons (winter and summer) have a significant impact in addition to the obvious impact factors (e.g. pipe material) of pipe on the failure rate, and the failure rate was almost two times higher for water mains and distribution conduits in winter than in the rest of the year in a case study of a Polish city (Kutyłowska and Hotloś, 2014).

Frost action lasts over a time period which corresponds to the period of sustained cold weather which results in increase in burst rate. The potential of sub-zero temperature to include these short term effects will depend on the moisture content of the soil and prevailing ground temperature, which will depend on recent (over a period of the previous weeks) historical metrological conditions.

Freezing Index (FI) was used as a surrogate for temperature effects on pipe breaks. The FI provides a measure of how severe a particular winter period was (Kleiner and Balvant, 2002). The periodical FI is expressed in degree-days, which is the cumulative daily mean temperature below a threshold temperature τ ($^\circ$C) during a given period.

$$FI_p = \sum_{\substack{i \in p \\ \forall T_i \leq \tau}} (\tau - T_i) \tag{3.1}$$

Where, FI_p is Freezing Index (in $^\circ$C-day) in period p, τ is threshold temperature (e.g. 0°C), and T_i is average daily temperature of day i.

(7) Corrosion and Other Environmental Indicators

Corrosion is one of the main causes of pipe recession. It is closely related to the physical characteristics of the pipe, the surrounding environment and maintenance management. The pipe material fundamentally determines the occurrence and development of corrosion but the surrounding environment (for example: external protection and soil conditions) has an impact on the process of corrosion.

Iron-based water main, which is a widely used pipe material, has inherently high structural strengths but is vulnerable to corrosion. Corrosion is one of the main causes leading to deterioration although the corrosion mechanism is different for different material.

In water transport and distribution system, corrosion process occurs and attacks pipes and joints. Corrosion is actually a series of related chemical and biological reactions. The common forms of pipe corrosion are galvanic corrosion, pitting, tuberculation, crevice corrosion, erosive corrosion, cavitation corrosion, biological corrosion.

Corrosion can be external corrosion, where the pipe wall reacts with its surrounding environment (namely the soils and its constituents), or internal corrosion, where the pipe wall reacts with the water flowing through it. Depending on the pipe material, there can be metallic corrosion and the corrosion of cement-based products. With metallic corrosion there are three types of reactions: hydrogen evolution (due to aggressive waters (with low pH)); oxygen reduction (occurs with normal waters); and sulphate reduction (occurs in anaerobic conditions occurring in soils). With cement-based (or lined) pipes (both concrete pipes and cement lined pipes), corrosion is the result of cement dissolving due to leaching of calcium at low pH. With concrete pipes this can result in metal reinforcements being exposed and damaged, affecting the strength and structural integrity of the pipe.

Water quality is a decisive factor affecting internal pipe corrosion. The main factors affecting the internal corrosion of metal pipes include pH, dissolved oxygen, temperature, flow velocity, turbulence, alkalinity, calcium hardness, chlorine and sulphate, chlorine residue etc.

Due to the comprehensive corrosion and corrosion prevention mechanism, the corrosion process, location and degree is difficult to be estimated accurately.

3.   Operational Indicators

Operational factors in Table 3.1 could have an effect on pipe deterioration, some of them act as causes and others act as results. Frequency of supplies, duration of water supplies and number of valves play some roles in intermittent water supply, which is beyond the scope of this research.

(1) Number of Connections

More connections or branches in a pipe lead to flow turbulence changes and the impact on the main pipeline. Moreover, connections or branches of a pipeline also give more opportunity to increase the leakage. Meanwhile, the change of flow pattern also increases the possibility of pipe break. Then, the number of connections impact can be represented indirectly by breakage history.

(2) Leakage Record

Leakage is one of the pipe deterioration results instead of causes and an indicator easy to be monitored. However, the leakage record also depends greatly on operation pressure rather than pipe cracks alone. Generally, leakage record is for part of a network or an entire network, instead of an individual pipe. Therefore, leakage rate is an indicator for the network it represented. The leakage rate for a specified pipe has to be inspected purposely.

(3) Complaint Frequency

Complaint frequency is also an indirect indicator to measure the pipe deterioration results. But these data reflect the water supply situation in a region, rather than a pipeline. In some cases, the causes of customers' complaints are complex and not necessarily pipe deterioration.

(4) Breakage History

Numerous studies have shown that breakage rate has an accelerated tendency after the first breakage. Pipe failures have been seen to occur in the vicinity and soon after previous bursts and repairs. This may be attributed to damage caused by the previous event or the disturbance caused by the previous repair. For example, it is found that beyond the first failure on a main, the number of failure events increase with time. Similarly, for mains with diameter greater than 200mm, the time to the next break decreased as each break occurred (e.g. Saegrov *et al.,* 1999; Kleiner *et al.,* 2001; Clark *et al,.* 2002; Sadiq *et al.* 2004). There could be many reasons for these occurrences including changes in soil moisture due to water from leaks and exposure to the extreme cold, and bedding disturbance during repair.

Although the overall mechanisms leading to pipe failure are often a combination of loading and structural deterioration, which are likely to be related to material, diameter, length and age, the macro performance show that future breakage rate has a good correlation with the historical breakage rate. Therefore, breakage history is an integrated pipe deterioration indicator, which should be paid attention to.

3.2.2 Influence Factor Selection

There are many factors affecting pipe deterioration, and the mechanism is complex. Most of the influencing factors analysed in engineering experience and research are summarized in Section 3.2.1. The selection of influence factors is quite different in different researches and different water supply companies. Therefore, it is the key to choose a suitable influence factor set. The study of Zangenehmadar and Moselhi (2016) aims to benefit from the Delphi method to prioritize the factors affecting the failure based on their significance. The reasonable selection of influence factor can effectively reduce the use cost of the index, improve the work efficiency, and get the accurate evaluation conclusion. If the item of the influence factors is more, it will increase the cost of assessment (Haider *et al.,* 2015). For mall water supply enterprises, a number of influence factors with greater impact must be selected due to limited funds, data, personnel, business capacity.

Clark *et al.* (2010) argued that while it is possible to identify the major factors that affect the reliability of a WDS, there are a number of unknown factors that affect pipe breakage rates. They present models for predicting pipe break rates and the cost-benefits associated with different inspection technology.

Because some influence factor values are difficult or impossible to obtain completely, no model could include them all. Spatial and temporal variability and randomness also result in complicated impact. Therefore, some significant and available factors are usually selected in pipe deterioration assessment. Among numerous influence factors, the principles of choosing a part of the factors to carry on the pipe condition assessment are:(1) the data that are available; (2) it generally has great effect on the pipe deterioration; and (3) it has a certain comprehensive factor so that it can represent some other factors that are not directly considered in pipe condition assessment.

Strictly speaking, the most significant factors to pipe deterioration can be identified if the data are available. However, for most water utilities, the available data are still limited. Boxall *et al.* (2007) stated that data available from most water companies in the UK and deemed to be of interest for the derivation of burst models are asset data and incident/customer service records. Asset data include pipe reference number, diameter, date laid, material, length, rehabilitation work carried out (including date of work), and date of abandonment. Incident service records include incident date, location, specific incident type, some comments of required action. Some data (e.g. weather record, road load record) are out of the water companies' database.

Kleidorfer *et al.*(2013) took sewer systems as objects of study and investigated four possible factors impacting sewer rehabilitation strategies: (1) sewer aging, including sewer deterioration models which are trying to predict the aging behaviour; (2) city development (i.e. increasing or shrinking population); (3) climate change; and (4) vulnerability and risk assessment of sewer collapse. From the point of view of the working conditions and the mechanism of pipe decay, there is no essential difference between the water distribution system and the sewer system.

In this study, the pipe material is considered as the most important parameter to deterioration. The reason is that corrosion and load/strength must exert influence to pipe deterioration through pipe material. Because material is a non-numerical parameter and it has multi-relationships with other indicators contributing to pipe deterioration, classifying pipes based on the material is a relatively easy way to reduce the differences.

Pipe material can be regarded as an interior deterioration caused while some environmental indicators can be regarded as exterior causes. These exterior causes can only exert influence through pipe material. Among the exterior causes, soil corrosivity, surface permeability and ground water condition can influence pipe corrosion through material corrosion resistance. Buried depth and traffic load will result in different impact load on pipes. If hydraulic pressure exceeds the maximum pressure limit, which is determined by material, pipe will break. Exterior corrosion is considered the major cause for reducing pipe stress resistance and is mainly determined by pipe material and the soil condition. The interior corrosion is determined by water quality and pipe material. The internal corrosion resistance results in the roughness change as well.

Although corrosion is an immediate and key cause affecting pipe deterioration, the corrosion occurrence and procedure is determined by many factors and most of them are usually unavailable for water utilities. It is almost impossible to accurately assess each pipe's corrosion degree, location and tendency based on these comprehensive relations among these factors. Soil type is a key factor and available for most water utilities, but the spatial variability of soil in a specified water distribution system is not very great. Therefore, the deterioration impact from soil corrosivity is thought to be homogeneous for the same material pipes in a specified water distribution system in this study.

Due to data limitation, from a practical sense, the involved influence factors are not too much. For these reasons, pipe material, pipe age (installation year), diameter, length, temperature (freezing index) and historical break record are chosen as the variables for the pipe condition assessment model in this research. One reason is they are really important to pipe failure and another is that they are usually available in most water utilities databases.

3.3 Methodologies

3.3.1 Pipe Breakage Number Prediction Modelling and Methods

Pipe breakage number prediction model describes the overall situation of pipe deterioration from the perspective of pipe groups. The model proposed in this research focuses on the deterioration tendency for relatively homogeneous pipe groups, where each group has similar deterioration causes and tendencies. After this is done, regression analysis performed in order to predict the total failure number for each of the pipe groups.

Some of the underlying objectives and features of the proposed model are presented below.

(1) Outputs of the model should be easily integrated into optimization models.

In some physically based models (e.g. Chukhin *et al.* 2014; Jesson *et al.* 2013)**,** pipe deterioration is often described by some micro but direct parameters, such as corrosion depth and width (or area) of cracks. It is easy to understand but difficult to be observed. Moreover, the model results are not suitable as inputs for optimal rehabilitation decision models.

In some statistical models (e.g. Scheidegger *et al,.* 2015), parameters such as breakage rate and the probability of breakage, are often used to describe pipe deterioration. However, the occurrence of pipe breakage is random in nature and hence these models are difficult to integrate into decision making.

In some data mining models (e.g. Yan 2006), some indexes (e.g. pipe condition index) are derived from deterioration indicators to estimate pipe deterioration degree. Although they are comprehensive, the indexes proposed are difficult to combine with a cost analysis.

In this research a model that combines a statistical model with a data mining model is developed. It is believed that this model overcomes several of the issues described above.

(2) Model should avoid subjective judgment.

Several previous researchers (e.g. Yan and Vairavamoorthy, 2003) apply methods that involve inputs and judgments from experts and decision makers. In practice, this judgment may not be reliable (if the experts have limited knowledge) and can introduce bias. The proposed model avoids subjective judgment.

(3) Using multiple non-linear regression it develops functions that relate pipe influence factors with breakage number.

A function is an effective approach to develop a relationship between pipe breakage occurrence important influence factors. For example, Berardi *et al.* (2008) used multivariate non-linear functions to integrate simple functions to describe the pipe break principles.

Multivariate non- linear functions are a feasible method if the errors are not great and the sampled amount is abundant. Whether this method is feasible or not can be tested only through error analysis.

In this thesis a multivariate non-linear regression is applied. It is an effective approach that can be applied for the breakage number estimation of pipes of similar characteristics. It employs an evolutionary polynomial regression to calculate coefficients that minimize the squared deviations of the observed pipe breakage rates points from the estimates.

3.3.2 Modelling Process and Computer Implementation

Pipe deterioration modelling process and the major steps are as shown in Figure 3.1.

The model consists of four modules: (1) data classification and aggregation module; (2) weighted nonlinear regression module; (3) model testing module; and (4) individual pipe condition assessment module. Following is the detailed introduction to these modules.

1.   Data Classification and Aggregation Module

An important prerequisite is that the pipes in a relatively homogeneous group have similar deterioration tendency or failure probability.   In order to eliminate breakage difference due to various influence factors, the pipe assets must be grouped according to homogeneity of influence factor (e.g. pipe material, pipe age (installation year) and diameter).

At the first, asset data are classified into some small homogenous groups (e.g. a pipe group with same pipe material, same pipe age and same diameter). However, such a detailed and fine classification might lead to some groups including too many pipes while some other groups including a few. Therefore, the fine classification groups have to be aggregated to some large groups. Data classification and aggregation in this model can be processed in Microsoft Excel. The main factors influencing pipe deterioration and their impacts have been narrated in Section 3.2. For non-numerical parameters (e.g. pipe material) and discrete parameters, they are the basis of data classification. Proper pipe cluster or classification according to non-numerical parameters can off-set some other influence in the same group.

The primary criterion is pipe material. All the research in the model is based on the premise of same material. The secondary criteria are pipe age (installation year) and diameter. If there are m groups for pipe diameter and n groups for pipe age, the total group number will be $m \times n$

theoretically. If pipe length is listed in an *m*-column and n-row table, each cell represents the sum of pipe length for a specified pipe group with the same age and diameter. Similarly, a statistical pipe breakage table can be obtained for a specified year. The pipe length in different classes may vary greatly and some groups might be null. Data aggregation will aggregate pipes in the same diameter or same age so that the data in any group are not so rare.

In order to include more influence parameters and more information in available data, this module has to overcome some drawbacks:

(1) The module should have the adaptability to include more parameters, rather than the two or three parameters in some existing models. Only if a parameter data is numerical and can be obtained, will it be included as input variable.

(2) The information fuse approach can be changed to reduce the randomness in some groups.

According to these requirements, formula selection and multiple nonlinear regressions will be carried out after data aggregation and classification.

2.   Weighted Nonlinear Regression Module

This module is to select a proper formula pattern, evaluate coefficient values through regression analysis, and provide static analysis results. The weights are derived from pipe length in each group.

In the existing literature, the functions between pipe failure number and the influence indicators (e.g., diameter and age) are often power function, exponent function or their mixture function (e.g. Constantine and Darroch, 1993; Mailhot *et al.*, 2003). Although these functions are the simplification of real nonlinear relations and none of a function pattern is perfect in all cases, the basic alternative formula patterns are the foundations of a further complicated pattern.

For a specified function type, the coefficient evaluation is a multiple nonlinear regression problem. Because it is obvious that the total pipe length in each groups are not the same, the pipe length of each group can play as weight in multiple nonlinear regression. Weighted nonlinear least square method is used to evaluate the coefficients in the formula. Because the sum of pipe length in each group is not the same, the observations errors are not equally reliable a weighted sum of error squares should be minimized.

3.   Model Testing Module

To verify the correctness of the model, part of the data can be used for modelling training and the other for model testing. Therefore, all the pipes are randomly divided into two parts, one is for coefficient evaluation and the other is for formula testing. This module uses part of the pipe data in the same network to test the accuracy of the formula. For example, use the observation data from Year 1 to Year $n-1$ to test the failure number in the Year n. If there is a good correlation and small errors between the estimated and recorded failure number, the methodology is proved to be feasible. Otherwise, the formula needs to be re fitted.

4.   Individual Pipe Condition Assessment Module

This module is to estimate a pipe's current structure condition by reinforcing its historical failure record based on the general failure rate of the group of pipes that it belongs to. According to the failure number prediction formula and its break history, an individual's nominal pipe failure rate is generated as the surrogate of pipe structure condition assessment indicator. It can be used for quantifying a pipe's structure condition. The historical breakage rate is adopted in the formula as the representative of unaccounted and random factors.

Start

Select the pipes with complete feature data and divide them into two groups randomly. The group with more pipes is used for model training, and the other is used for model testing.

Module 1: Data Classification and Aggregation

Begin

Inputs: asset data (e.g. pipe diameter and age) and pipe failure (break) data of each selected pipe

Classify the pipes by the homogeneous features (e.g. same pipe material and same age)

Aggregate data by some relative homogeneous features (e.g. same pipe material, or same age)

Outputs: asset data groups with homogeneous features

End

Module 2: Weighted Nonlinear Regression Module

Begin

Inputs: asset data groups with homogeneous features

Establish some possible formula types by assumption and experience

Select a formula and evaluate the coefficients by using multiple nonlinear regression and the sum of pipe length in a group treated as weights

Outputs: the most possible formula of pipe break prediction

End

(To be continued)

Module 3: Model Testing

Begin

    Inputs: the failure and asset data for testing after classification and aggregation

    Hypothesis testing: the errors between the prediction and record are acceptable or not

      If *Yes*, accept the approach and further derive the formula by using all the available data

      If *Not,* find the possible reasons or non-applicable cases leading to the methodology failure

      End

    Confirm formula and predicted pipe break number for a homogeneous pipe group

    Outputs: the confirmed formula of pipe breakage prediction or the reason leading to the methodology failure

End

Module 4:   Individual Pipe Condition Assessment

Begin

    Inputs: an individual pipe's break history record and other parameters

    Assess individual pipe's structure condition by nominal breakage rate

    Outputs: individual pipe's condition

End

End

Figure 3.1   Pseudo code for pipe deterioration assessment

The needed data includes water distribution asset data, pipe failure (break or repair) data and temperature data. Table 3.2 lists each module's goal, input and output.

Table 3.2   Module descriptions for pipe condition assessment model

No.	Module	Goal	Input	Output
1	Data classification and aggregation	To stress the common feature's influence to the pipe deterioration	Raw asset and failure data	Classified data with homogeneous features
2	Weighted nonlinear regression	To search for a suitable function type and coefficients value	Aggregated failure and asset data	Failure number estimation function
3	Model testing	To test the feasibility and accuracy of the regression module	Other independent failure and asset data	The errors and correlation coefficient between the prediction and failure record
4	Pipe condition assessment	To assess individual's structure integrity	Individual's information (e.g., break record)	Nominal breakage rate

3.3.3 Multivariable Nonlinear Regression Analysis

In this model, the main task is to evaluate the values of coefficients in a nonlinear function. Nonlinear regression is where the observational data is modelled using a nonlinear function to combine the model parameters. The curve of best-fit can be generated by minimising the sum of squared residuals.

For a non-linear function, the linear transformation may be a useful approach but such method also deforms the errors. If the function cannot be linearly transformed or the errors are not to be deformed, non-linear least squares method is to be applied.

1.   Nonlinear Least Squares

The nonlinear least squares formulation is used to fit a nonlinear model to data. A nonlinear model is defined as an equation that is nonlinear in the variables, or a combination of linear and nonlinear in the variables.

Consider a set of m data points, (x_1,y_1), (x_2,y_2), …, (x_m,y_m), a model function curve $y = f(x,b)$, with variables x and n parameters in the function, and coefficients vector $\vec{b}=(b_1,b_2,... , b_n)$, with m≥n. It is desired to find the coefficients vector $\vec{b}$. Hence, the least square method is used to fit the curve so that the error sum of the curve and the given data is the minimum.

The residual between observational data and fitted value is defined as:

$$r_i = y_i - \hat{y}_i \qquad (3.2)$$

Where, r_i is the i-th residual, y_i is the observation value, and $\hat{y}_i$ is the fitted value.

The summed square of residuals is given by

$$S = \sum_{i=1}^{n} r_i^2 = \sum_{i=1}^{n} (y_i - \hat{y}_i)^2 \qquad (3.3)$$

Where, n is the number of data points included in the fitting and S is the sum of residual squares estimation.

Nonlinear models are more difficult to fit than linear models as they require an iterative approach to estimate the coefficients:

(1) Start with an initial estimate for each coefficient (this can be done randomly or using some form of heuristics).
(2) Produce the fitted curve for the current set of coefficients. The fitted response value $\hat{y}$ is given by $\hat{y} = f(X,b)$ and involves the calculation of the Jacobian of $f(X,b)$, which is defined as a matrix of partial derivatives taken with respect to the coefficients.
(3) Check to see if the fit is acceptable. If so stop, otherwise continue to step (4).
(4) Using an appropriate algorithm, adjust the coefficients in a systematic way and return to Step (2). Example algorithms for adjusting coefficients include Trust-region, Levenberg-Marquardt and Gauss-Newton).

2. Weighted Nonlinear Least Squares

It is noted that weights can be used for nonlinear models, and the fitting process is modified accordingly. In the weighted non-linear least squares, weighted errors distinguish the different reliability of group errors. Moreover, the weights assignment depends on the pipe length in each group, rather than decision maker's subjective bias and judgement in pipe condition assessment.

Because the pipe length of each group is different, the contribution of breakage number residuals is different in the error analysis. The greater the sum of pipe length is, the greater the weight. Weighted least squares regression can be used where an additional scale factor (the weight) is included in the fitting process. Weighted least squares regression minimizes the error estimate.

$$S = \sum_{i=1}^{n} w_i (y_i - \hat{y}_i)^2 \tag{3.4}$$

Where, S is the weighted residual sum of squares, w_i is the weight, y_i is the observed value and $\hat{y}_i$ is the calculated value by the function. The weights determine how much each response value influences the final parameter estimates. Weighting the data is recommended if the weights are known, or if there is justification that they follow a particular form.

3.4 Modelling

The pipe breakage number prediction model is corresponding from Module 1 to Module 3. Individual pipe condition assessment is in Module 4. The components of pipe deterioration assessment and relationships between each module are listed in Figure 3.2. The major modelling steps are as follows:

(1) Pipe classification and aggregation: To classify and aggregate asset and breakage data
(2) Regression analysis (weighted multiple nonlinear regressions): To select part of the pipes' data randomly to fit a proper formula to predict a specified year's breakage number through regression analysis.
(3) Model test and formula fitting: To test and validate the prediction with the rest of pipes data through hypothesis testing. This is to confirm that the method is feasible. If hypothesis testing is passed, then, the next step is to fit a proper formula by using entire network's pipe data. This is to find a more accurate formula by involving more information. If hypothesis testing is a failure, it might be due to data shortage or some unconsidered factors.
(4) Individual pipe condition assessment: To assess individual pipe condition

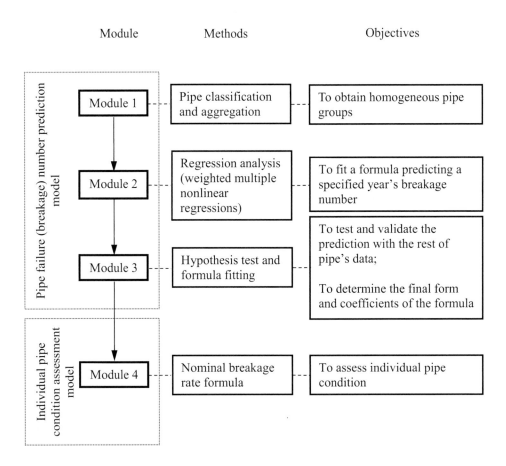

Figure 3.2   The components of pipe deterioration assessment

3.4.1 Pipe Classification and Aggregation

Pipe material is a typical non-numerical parameter and also one of the most important factors determining pipe deterioration. Therefore, all the pipes are clustered according to their material at first. Then, pipe diameter and age are taken as the sub-criteria of classification after material. If more data are available, the clustering can be done further by the other homogeneous parameters. However, such an ideal method meets difficulty in practice. Even if with a large database comprising tens of thousands of kilometres of main, at some point, the pipe groups and the number of burst events corresponding to these pipe groups become so small that the analysis usually collapses (Skipworth 2002). For this reason, proper aggregation will be carried out after classification so that the failure events in a group are not too few.

Under the premise of same pipe material, aggregation is a relatively rough classification which only considers one homogeneous character, rather than two or even more at the same time. For example, consider same diameter or same pipe age respectively.

Since one of the research objectives is the failure number estimation for pipe groups, such estimation mainly depends on the certain or general deterioration tendency among the pipes rather than some casual or random factors. The individual pipe's difference is temporally ignored in the same group. The pipe aggregations reinforce the common factor's influence and weaken the influence from individual pipe's different characters simultaneously. For example, a group of pipes with the same material, diameter and age can be assumed to be jointed one by one as an imaginary long pipe. These pipes' overall deterioration process is only affected by the common factors. All the failures that occur in such an imaginary long pipe can be thought to being driven by the common influence factors (Berardi *et al.* 2008). Therefore, the failure number difference between groups is considered mainly driven by the same parameters (e.g., difference of diameter). The exact failure timing and location is random due to some unaccounted for factors and inherent randomness.

An individual pipe, usually tens or hundreds meters in length, has rare breakage before replacement. Pipe breakage is almost a random accident for a pipe during a relatively short observation time (e.g. a few years) because the occurrence of pipe break is a discrete and small probability event. Such randomness or 'noise' almost dominate the break occurrence on a relatively short pipe. However, with the sum of pipe length increase, the occurrence of breaks becomes more frequent. The regularity of statistics is more obvious, and the effect of pure randomness is less evident (Kleiner *et al.*, 2007).

An assumption in the method is that the other unaccounted for influential factors remain evenly distributed across the groups when the asset database is clustered into homogenous pipe groups by a parameter. Actually, the influence from other ignored factors can also be regarded as off-set or weakened within such a group.

Through data aggregation, more failure records are accumulated in a relatively homogeneous group so that the main influence factors are strengthened. All the features in a group are expressed by equivalent parameters, which are defined by the weighted average of each actual parameter value. The pipe lengths act as the weights (Berardi *et al.* 2008). The equivalent diameter, age and length are as follows:

$$Age_{class} = \frac{\sum\limits_{i=1}^{N}(L_i \cdot Age_i)}{\sum\limits_{i=1}^{N}L_i} \tag{3.5}$$

$$D_{class} = \frac{\sum\limits_{i=1}^{N}(L_i \cdot D_i)}{\sum\limits_{i=1}^{N}L_i} \tag{3.6}$$

$$L_{class} = \sum\limits_{i=1}^{N}L_i \tag{3.7}$$

$$FI_{class} = \frac{\sum\limits_{j=1}^{T}(FI_j)}{T} \tag{3.8}$$

$$Br_{class}^{re} = \frac{BR_{class}^{re}}{\sum\limits_{i=1}^{N}L_iT_i} \tag{3.9}$$

Where, Age_{class}, D_{class}, L_{class} and FI_{class} are equivalent pipe age, equivalent diameter, equivalent length and equivalent freezing index for a group respectively, the subscript of *class* means the class or equivalent value, L_i, Age_i, and D_i are the *i-th* pipe length, age and diameter respectively in this group, FI_j is the freezing index (FI) in the $j-th$ year during the monitoring time horizon. The definition of FI has been introduced in Section 3.2. The total failure data covers T years and the total pipe number is N. The observation period is T years as well. Br_{class}^{re} is the average historical breakage rate of this pipe group in the observation period, BR_{class}^{re} is the total recorded (historical) breakage number of this pipe group in the same observation period.

Since the pipe data classification makes failure data sparse and uneven among these basic groups with the same material, diameter and age, further aggregation that accumulates more assets and failure data are necessary. The principle of aggregation is that all of the assets and their failure data are grouped either by the same diameter or by the same age in the case of same material in spite of other characteristics. For each aggregated group, the common characters are represented by the equivalent parameters. Through the aggregation, more asset and failure data are accumulated in a group.

3.4.2 Regression Analysis

In this research, part of the asset data and their failure record are used for model training and the residual data are used for model testing. If the function derived from training can well predict the failure number for the testing data, it proves the feasibility of this method.

As previously described, in addition to pipe material, five indicators are chosen as the independent variables in this model. They are pipe age, length, historical breakage rate, temperature (freezing index) and diameter. The research task is to find a reasonable function type and estimate the coefficient values through regression.

1.   Function Type

There is no authorized or universal function type suitable to water main break number prediction of any WDS. One reason is the diversified influential factors. The other reason is that the complicated deterioration mechanism is still not understood completely. One more reason is that the inherent randomness of burst always leads to some uncertain results. Meanwhile, data shortage also brings some difficulties.

In the existing researches, almost all of the function forms are based on some assumptions rather than some convictive theory. Although the function forms are infinitive in theory, some literatures have proposed some feasible and approximate optimal function types according to their data. In the existing literatures, exponential function, power function, logarithm function and their combination are most frequently used formulae in existing research. For example, exponential function is adopted in numerous researches (e.g. Mavin, 1996). Constantine *et al.* (1996) utilized a power function. Mavin (1996) compared the time-exponential model to a time-power model and found that the performance of the two models in predicting water main breaks was comparable. Boxall *et al.* (2007) used combination of power function and logarithm function to address annual burst rate which can be converted into actual burst rate by evaluating the exponent. Generally, these alternative function types are not definitely suitable for any networks, but they provide some alternative function types at least.

If a breakage number or breakage rate is expressed by a determinate function, it shouldn't be thought that this is a determined event. On the contrary, it should be understood as the mathematical expectation or the most likely value in the estimation. There must be a great error if the breakage number and breakage rate are measured at an individual pipe's level.

Based on existing literature (e.g., Berardi *et al.* 2008), an undetermined multivariable function combining exponential and power functions is proposed in our study:

$$Y = a_0 \cdot \prod_{i=1}^{n} f_i(x_i) \tag{3.10}$$

Where, a_0 is an undetermined coefficient, x_i is deterioration influential factors (e.g. diameter), $f_i(x_i)$ can be a power function or exponential function of a single variable, i.e. $f_i(x_i) = x_i^{a_i}$ or $f_i(x_i) = \exp(a_i \cdot x_i)$, a_i is coefficient for each parameter. These types are the conversion and combination of simple power function or exponential function.

It is assumed that the total break number linearly increase with the pipe length in the same pipe group. Because the pipes in the same group are assumed to have homogeneous features, the breakage rate (break number per kilometre per year) has no relation to the length. Hence, there are only four undetermined relationships between break number and the other four indicators.

According to common knowledge, break number generally has the same change tendency with pipe length, age, freezing index and historical break record increasing, and the opposite change tendency with the diameter. Such a tendency can determine whether the coefficient a_i in $f_i(x_i)$ is positive or negative. For the former four indicators, the coefficients should be positive, while for diameter, the coefficient should be negative. Using the common knowledge will reduce the probability of wrong judgment in regression and narrow the searching scope.

Since each indicator has two possible sub-function types and for that length is determined, there are only $2^4 = 16$ alternative function types. The general function type will be:

$$BR_{class}^{es,t} = a_0 \cdot L_{class} \cdot f_1(Age_{class}) \cdot f_2(FI_{class}) \cdot f_3(Br_{class}^{re}) \cdot f_4(D_{class}) \tag{3.11}$$

$$Br_{class}^{es,t} = \frac{BR_{class}^{es,t}}{L_{class}} = a_0 \cdot f_1(Age_{class}) \cdot f_2(FI_{class}) \cdot f_3(Br_{class}^{re}) \cdot f_4(D_{class}) \qquad (3.\ 12)$$

Where; the superscript of es means the estimated values, the superscript of re means the recorded values, the superscript of t means the specified year t , the subscript of $class$ means the equivalent value, BR is the breakage number, Br is the breakage rate. For each alternative function type, five coefficient (i.e. a_i (i=0~4)) values needed to be determined.

2. Coefficient Evaluation

For any assumed function type, weighted nonlinear least squares method is used for coefficient estimation. The primary objective is:

$$\text{Min } \sum_{k=1}^{NG} (BR_{class}^{es} - BR_{class}^{re})^2 \qquad (3.\ 13)$$

Where, NG is the total number of pipe group, BR_{class}^{es} and BR_{class}^{re} are estimated and recorded break numbers of the class respectively. BR_{class}^{es} is calculated through Eq. (3. 11), and BR_{class}^{re} is collected from failure record database. It must be noted that each group pipe has a different length, which means the error for each group of data has different weights (or contribution) to the total error. The objective will be modified by adding the weights:

$$\text{Min } \sum_{k=1}^{NG} w_k (BR_{class}^{es} - BR_{class}^{re})^2 \qquad (3.\ 14)$$

The primitive concept of the weigh here refers to the ratio of the sum of a group of specific pipes' length to the total length of the chosen pipes. If some of the pipes are installed during the observation period and the service time is shorter than some other pipes, the definition of weights should be determined by "service length" which is the product of pipe length and its service time span during the observation period.

$$w_k = \frac{\sum_{i \in k}(L_i \cdot T_i)}{\sum_{k=1}^{NG} \sum_{i \in k}(L_i \cdot T_i)} \qquad (3.\ 15)$$

Where, w_k is the error weight of the $k-th$ pipe group, T_i is the service time span, $i \in k$

means in the case of that the $i-th$ pipe is an element of the set of the $k-th$ pipe group, $(L_i \cdot T_i)$ is the $i-th$ pipe's service length, NG is the total number of pipe group.

This is a weighted multiple nonlinear regression model. Eq. (3. 14) can be converted into a general Least Square objective function:

$$\text{Min} \sum_{k=1}^{NG} (\sqrt{w_k} BR_{class}^{es} - \sqrt{w_k} BR_{class}^{re})^2 \tag{3. 16}$$

It seems that Eq. (3. 11) can be easily converted into a linear function by a suitable transformation of the model formulation. If a logarithm conversion is taken on both sides, the function becomes linear. However, such a nonlinear transformation can change the data value, the error structure of the model and the interpretation of any inferential results. Hence, the linear transformation is not adopted in this model.

There are some methods for nonlinear regression, such as Gauss-Newton method, Newton-Raphson method, Levenberg-Marquardt method etc. Meanwhile, there are also some software to solve a nonlinear regression problem, such as Matlab, OriginPro, SAS, SPSS, DataFit and GraphPad. In most of the cases, the initial values are required. However, it is also difficult to provide or guess the initial values unless there is some experience. In this research, software called "*1stopt*" (i.e., "First Optimization"), which was developed by 7D-Soft High Technology Inc., is applied for nonlinear regression to evaluate the coefficients value. The advantage of this software is its global searching capability. By this software, the initial values are given by the computer randomly instead of being required by analysts and it can provide correct answers in most of the cases. Algorithm selection and iteration numbers are selected by analysts which control the accuracy and speed of calculation. The combination of Levenberg-Marquardt and Universal Global Optimization, which is applied in our study, is one of the suggested algorithms by this software. Except for coefficient values, the error and related coefficient can be derived through this software. Among all the function type alternatives, one of them with higher related coefficient and low weighted errors will be chosen as the suitable function type. Furthermore, if an assumed function type can be tested and proved to be accurate enough in specified water distribution system rehabilitation's case study, such a methodology can be confirmed and this function may be applied in the same water distribution system.

3.4.3 Model Test and Formula Fitting

1.   Model Test

The function and the coefficients are derived from part of the asset and failure data. Therefore, its validity should be tested and verified by other independent data in the same water distribution system. Since the model is to predict the pipe failure number in a specified year by assuming it follows a general deterioration tendency, the difference between prediction and observation should not be significant. Nevertheless, this must be tested.

In the test, equation type and coefficients are the same as these in regression analysis. Different historical period data are used to predict the breakage number in the next year. For example, prediction and testing are based on ten year observation records. In the regression analysis, some pipes' date in the earlier nine years are used as historical data to predict the last one year's breakage number.

It is assumed that there is not significantly difference between estimated and recorded breakage rate. However, this proposition must be confirmed though a hypothesis test. Because the estimated and recorded breakage data come in pairs, the difference of a pair can be regarded as the residual. Therefore, the differences in these different pairs follow the same probability distribution.

Suppose there are n pairs of breakage number data (e.g., (BR_1^{es}, BR_1^{re}),..., (BR_n^{es}, BR_n^{re})). The corresponding error data are $error_1 = (BR_1^{es} - BR_1^{re})/L_{class1}$, ..., $error_n = (BR_n^{es} - BR_n^{re})/L_{classn}$, which are independent to each other. After weighting, these weighted error data constitute a sample of a normally distributed population. It is assumed that $error_i \sim N(\mu, \sigma^2)$, i=1, 2, ...,n, both the expectation and variance are unknown. A hypothesis can be addressed as:

$$H_0 : \mu = 0, \ H_1 : \mu \neq 0$$

If the hypothesis is accepted, this means the breakage number prediction and observation has no significant difference. Then the formula can be used in the same network and same predicted year. If the hypothesis is not accepted, this means the prediction formula cannot be used due to the significant difference. Generally, the situation can be classified into two cases:

(1) The errors are large in the same year in regression analysis. There are two possible causes. One is that the unaccounted for influence factors are quite different for these two groups. These unconsidered factors might control the failure occurrence and breakage record. Another possible cause is that the testing data amount is not large enough and the randomness dominates the failure. Generally, the latter reason can be avoided easily by adding some testing data before testing but the former is not predictable. If it really happens, the dominated unconsidered factors must be discovered and accounted for in the formula. The type of the function might also be modified as one or more indicators are put in as exponent or power term. According to literature, the unconsidered factors having great impact on deterioration are most likely to be soil corrosiveness, pipe protection, workmanship etc.

(2) The errors are large in the different year from regression analysis. When the model is extended into failure prediction in other years, the error might be large in testing. One of the reasons is that the actual dynamic factors (e.g. freezing index) values in the estimation year might be quite different from the historical average values. Another reason might be that the randomness or unaccounted for indicators dominate the occurrence in the case if the observation horizon is short.

Although good correlation and small errors can confirm the method's function's feasibility, the testing errors also provide some information and solution to modify the formula. Long term observation data is the essential solution to reduce the influence from randomness.

2.   Formula Fitting

The more abundant the historical record and longer the observation time are, the less effect from randomness, and less error is in the model testing. Therewithal, the correlation coefficient and the total errors are closely related to the length of observation time. Based on previous analysis, this formula generating approach is valid if the hypothesis is accepted. The next step is to utilize entire networks and entire observation window's data to fit a suitable formula for the whole network.

It should be noted that the formula type and coefficients can be updated with new data. If the formula is used to predict breakage number in the near future (e.g. next year), the error is generally acceptable if the hypothesis is accepted. However, the errors of the far future become larger. The main reason is that there is some data gap between the current and the future, and the gap increases with the time prediction span. The typical cases are recorded (or actual) breakage number and freezing index.

Actual breakage record of the following years is unknown so that the historical breakage for the far future is not accurate any longer. One possible solution is that the predicted breakage number of a specified group in a year of the future is used as a surrogate of actual breakage number year by year, but the errors become larger. Another possible solution is always taking the historical breakage records as fixed values for a specified pipe group. The latter is easier for calculation. For these reasons, the data and formula should be updated frequently by adding new data.

3.5 Summary

Although there are some non-destructive pipe detective approaches, they are still time and cost consuming in dealing with large quantities of water distribution network pipes. This chapter develops an innovative pipe breakage number prediction model to estimate a group of pipes' breakage numbers in a specified year. Only six important and data available parameters (i.e., pipe material, pipe age, pipe diameter, pipe length, historical breakage record and freezing index) are chosen as the main influence factors although there are numerous factors influencing deterioration. Such simplification makes the method focus on some key factors and becomes feasible.

Through literature review, it is found that no existing model is perfect. As an input of further criticality assessment and optimization model, a crisp and quantified indicator to address pipe deterioration is necessary. This indicator is breakage number of a specified group of pipes. Some explicit function is needed to bridge the various influence indicators and breakage number that represents the general deterioration tendency. The pipe breakage estimation for a group of pipes denotes the most likely breakage number for these pipes in a specified year.

In the modelling, assets and their failure data are firstly classified with some relatively homogenous features, which are pipe material, age and diameter. The further aggregation with equivalent parameters extends the classifications. The aggregation accumulates more data and strengthens the pipe failure causes due to common features represented by equivalent parameters. The influence of a single factor can be emphasized in such a method.

The primary pipe breakage number estimation formula type selection is limited within some typical forms according to existing research but the coefficients are fitted through regression. The formula with good fitness was chosen to predict the failure number for a group of pipes. Asset data are classified into two groups, one is for training and the other is for testing. The formula is derived from training data and also tested by testing data in the same water distribution network which are assumed to follow a similar deterioration tendency. If the breakage number prediction and observation has no significant difference, the methodology and formula is feasible. Otherwise, more observation data should be input into the model.

The methodology in the model can be extended to involve more influence indicators and be applied in other pipe material as well. If more data are available (i.e. more monitoring horizon and more pipes), the confidence of the regression results will increase. In order to increase the models' reliability, these models should be routinely revised every year to update deterioration rates according to recent breakage patterns.

Chapter 4   Pipe Criticality Assessment Model

4.1 Introduction

Pipe structural condition is not the only attribute to decide whether a pipe should be renewed or not. A pipes hydraulic significance within a distribution network is also an important factor (Yoo *et al.*, 2014). In this chapter a pipe criticality assessment model is presented that combines an estimate of the condition of individual pipes, with their hydraulic significance, to generate a criticality index for individual pipes.

This model aims to reduce the dimension of the whole-life cost pipe rehabilitation optimisation model that is considered in the next chapter. It does this by identifying a subset of key pipes for the optimisation to consider and hence reduces search space for the optimisation process. This will reduce significantly the computational effort expended during the optimisation process. Hence the criticality model presented in this chapter, identifies the priority pipes for which action should be taken.

Although the quantification of pipe criticality assessment is important, there is no authorized indicator or measurement to quantify it. Even for the term 'criticality', there are different interpretations. For example, criticality can mean the consequence of a failure of a system component (e.g. Lippai and Wright, 2005). This thought considers the significance of the pipe without considering the structural condition of the pipe. In this thesis the term criticality relates to two components; pipe structural condition; and pipe significance. These two components are independent and hence it is necessary to propose a method to combine these two components into a single indicator that can be widely used to measure each pipe's criticality among hundreds or thousands of water mains in a WDS.

Combining the two components of criticality into a single indicator can be viewed as a Multiple Criteria Decision Making (MCDM) problem and there are many methods available to deal with this kind of problem. The selection of an appropriate MCDM method was a key task of this research work. In this thesis a modified TOPSIS approach is proposed. Although

TOPSIS is an effective MCDM method, the rank calculation is not always consistent with the TOPSIS principle completely. Hence, an innovative pipe criticality assessment model based on a modified TOPSIS is proposed. Its application in rehabilitation decision selection is introduced as well

4.2 Indicators Concerning Criticality

Pipe criticality is a comprehensive indicator describing the urgency (or priority) of a pipe to be renewed.

Relative criticality index (RCI) of pipes, integrates the effects of reliability, pipe breakage repair costs, and the energy demand to repair breaks in the pipelines (Piratla and Ariaratnam, 2011). RCI is comprehensive, but the requirement of the data is still high.

Ennaouri and Fuamba (2013) identified a set of 15 factors that affect pipe degradation in terms of hydraulic and structural aspects and determined the relative importance of these factors using the analytic hierarchy process (AHP).

Diao *et al.* (2014), developed a method that divides a WDS into clusters where components in the cluster have stronger internal connections than external one. Their method allows the identification of groups of critical components, and provides criticality prioritizations based on the rank of the groups.

Failure probability corresponds to pipe deterioration or structural condition, while failure consequence (cost) corresponds to pipe significance. A pipe with poor condition usually has high failure probability. Meanwhile, a significant pipe often results in more loss if pipe failure occurs. Since failure risk is difficult to assess owing to the difficulties in failure loss (or cost) and probability assessment, such an object can be converted into pipe criticality assessment in this study.

In this research, the implication of pipe criticality is relatively simplified, which is a combination of probability of failure and its consequence of failure. Therefore, pipe criticality and risk of failure are essentially consistent concepts though their definitions and measurements are apparently different. Pipe criticality is mainly determined by pipe condition and significance. Pipe condition is consistent with pipe failure probability. For example, a pipe with poor structural condition often has higher probability of failure. Meanwhile, pipe significance has some consistence with pipe failure consequence. Namely, a pipe with more significance usually results in more failure cost and more serious failure consequence. For example, a pipe's failure close to a water source would have greater impact on network's hydraulic performance than a pipe failure that is far away from a water source.

Criticality Index (CI), the measurement of pipe criticality assessment, is an index that combines pipe structural condition and pipe hydraulic significance (Figure 4.1). Pipe condition reflects a pipe's structural condition and significance reflects the hydraulic impact.

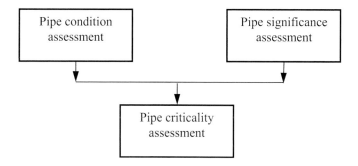

Figure 4.1   Concept framework of pipe criticality assessment

While quantifying the two factors that determine the criticality assessment, each pipe's potential renewal scheme is delivered as well. If pipe condition assessment is more prominent than pipe significance, the pipe should be replaced by a new one. Otherwise, the pipe needs a larger diameter or the pipe should be relined.

4.2.1 Pipe Condition Assessment Model

In order to simplify the terminology, the terms of pipe failure, breakage and burst are used interchangeably in this research although they are not exactly the same.

1.   Nominal Breakage Rate

The term of nominal breakage rate is a surrogate of pipe structure condition, which is a modified breakage rate that is associated with the individual's historical break record. Because the historical break record involves some implicit information and influence from both general deterioration tendency and the individual pipe's special features, this term is taken as an integrated indicator to affect the individual's condition assessment.

Historical breakage rate alone is not a suitable parameter for pipe condition assessment. In water utility's database, only a small number of water mains have break records. For most of mains, their failure records are zeroes. One reason is the record history is short so that the records are not complete. Another reason is the number of failure events is still much less than that of pipes in every year. Without other factors involved, it is almost impossible to tell the deterioration difference between two pipes with same accounted features (i.e., material, age, diameter and break history). However, this does not mean that their deterioration degree is same to each other. If the randomness is considered, this historical breakage rate cannot be the measurement of pipe condition.

It can be thought that pipe aging is a general tendency on each pipe, but pipe break is a representative phenomenon after deterioration which occurs on the vulnerable water mains. Since the pipes in a group are assumed to be relatively homogenous, the probability of break is evenly distributed in every unit length of all the pipes in the same group in previous model. This is under the assumption that the break is only determined by randomness factors. However, the real break occurrence is also affected by some unaccounted for factors. Since pipe deterioration mechanism is not described in this model, the historical failure record can be regarded as a comprehensive indicator for all of randomness and unconsidered influence factors. For such a reason, it can be an important factor to distinguish pipe's deterioration degree in a homogeneous pipe group. Under the impact of randomness, all of pipe sections in a homogeneous group have same probability of break because all of them are thought to be undifferentiated.

It is a common phenomenon that one pipe has breakage record while others in the same homogeneous group haven't any records. A probable reason is the pipe has an inherent flaw, or bad environment or disadvantageous operation condition, compare to other pipe in the same group but these factors are not accounted clearly. Although the factors causing breakage are complex, pipes with breakage history are more likely to break in the future than a pipe without. This phenomenon is partly due to the fragility of the pipe itself and the severity of the surrounding environment. For pipes with break history, this history should be emphasized to distinguish its condition difference from those without. Their particular breakage history should be integrated into the pipe condition formula.

A new term, "nominal breakage rate", involving a pipe's breakage record and the pipe group's general deterioration characters, is proposed as the surrogate of a pipe's condition.

The term "nominal breakage rate" can distinguish the pipe condition between two ideal exactly same pipes except for their breakage record. The one with more breakage number is assessed to have a higher nominal breakage rate than the other one.

(1) For the pipes without breakage record, all the pipes share the total break number evenly in a group. Then, the $i-th$ pipe's nominal breakage rate is the average estimated breakage of the group:

$$\hat{Br}_i^{es} = \frac{BR_{class}^{es,t}}{L_{class}} = a_0 \cdot f_1(Age_{class}) \cdot f_2(FI_{class}) \cdot f_3(Br_{class}^{re}) \cdot f_4(D_{class}) \qquad (4.1)$$

Where, Br_{class}^{re} is the recorded average breakage rate of a group, $\hat{Br}_i^{es}$ is the $i-th$ pipe's estimated breakage rate, $BR_{class}^{es,t}$ is the estimated total break number in the Year t.

(2) For the pipe with break record, its breakage number is added to the total breakage record of the pipe group so as to emphasis breakage records' influence on the specified pipe's condition assessment. Therefore, the $i-th$ pipe's nominal breakage rate is:

$$\hat{Br}_i^{es} = a_0 \cdot f_1(Age_{class}) \cdot f_2(FI_{class}) \cdot f_3(\hat{Br}_i^{re}) \cdot f_4(D_{class}) \qquad (4.2)$$

$$\hat{Br}_i^{re} = \frac{BR_{class}^{re} + BR_i^{re}}{\sum_{i=1}^{NP} L_i T_i} \qquad (4.3)$$

Where, $\hat{Br}_i^{es}$ is the $i-th$ pipe's estimated nominal breakage rate, $\hat{Br}_i^{re}$ is the $i-th$ pipe's recorded nominal breakage rate, BR_{class}^{re} is the recorded breakage number of a group, BR_i^{re} is the recorded breakage number of the $i-th$ pipe.

If the $i-th$ pipe's recorded nominal breakage rate $\hat{Br}_i^{re}$ replaces the recorded breakage rate Br_{class}^{re} in Eq. (3.11) and Eq. (3.12), Eq. 4.2 is obtained. The definition formula of $\hat{Br}_i^{re}$ is Eq. (4.3). The individual pipe's recorded breakage number is added to the total breakage number of the group to emphasis its particular evidence to assess the specific pipe's condition.

Although the nominal breakage rate has the same unit as breakage rate, it is only a surrogate of pipe structural condition. It cannot be used for real pipe break number calculation. The total estimated breakage number does not equal to the sum of the individual pipe's breakage number derived by nominal breakage rate.

If a pipe has a higher nominal breakage rate, it denotes worse structure condition. Apparently, pipe condition in the future can be predicted, but the reliability of prediction decrease over time from now on. Because the historical breakage records are updated every year, the coefficients in Eq. (4. 2) and historical average breakage rate should be changed as well.

2. Weighted Average

Since a pipe can be classified into different categories according to different criterion, its nominal breakage rate can also be derived from these different pipe groups, such as same diameter or same age. Generally, pipe's nominal breakage rates of different pipe groups are different. It is necessary to combine the results from different sources into one. Weighted average can be applied to generate a more believable conclusion.

With the premise of same pipe material, a pipe may belong to three pipe groups. They are the same diameter group, same age group and the same both diameter and age group. The third one is the intersection of the former two. If other criterion is applied, the classification can be further processed. Pipes' uniformity in a group increases with more criteria applied, but data amount (total pipe length) decreases so that randomness becomes significant. For each pipe group, there is always an estimated failure number if equivalent coefficients (e.g. equivalent diameter and equivalent age) are applied.

The ratio of an individual pipe length to the total length (equivalent length) of a group of pipes can be regarded as the individual pipe's influence capacity to the whole group.

$$r_{i,j}^l = \frac{l_i}{L_{class,j}} \tag{4. 4}$$

Where, $r_{i,j}^l$ is the relative length of the $i-th$ pipe in the $j-th$ group, l_i is the $i-th$ pipe length, $L_{class,j}$ is the equivalent (total) length of the $j-th$ group. If a pipe's length is small or the ratio is small, a pipe's individual feature has less impact on or less contribution to the group's general deterioration tendency.

Because one pipe may belong to a few different homogenous categories, it has different similarity degrees in these categories. This concept can be addressed by weights of similarity:

$$w_{i,j} = \frac{r_{i,j}^l}{\sum_{j=1}^{N} r_{i,j}^l} = \frac{\dfrac{l_i}{L_{class,j}}}{\sum_{j=1}^{N} \dfrac{l_i}{L_{class,j}}} \tag{4.5}$$

Where, $w_{i,j}$ is the $i-th$ pipe's weight in the $j-th$ group, N is the total number of pipe group. Therefore, the nominal pipe breakage rate is the weighted average values of these derived from different groups.

$$\hat{Br}_i = \sum_{j=1}^{N} w_{i,j} \cdot \hat{Br}_{i,j}^{es} \tag{4.6}$$

Where, $\hat{Br}_{i,j}^{es}$ is the nominal break rate for the $i-th$ pipe in the $j-th$ group, $\hat{Br}_i$ is the weighted average of the nominal break rate for the $i-th$ pipe. $\hat{Br}_i$ is used as individual pipe's condition assessment indicator.

4.2.2 Pipe Significance Assessment Model

The term of pipe significance denotes a pipe's failure influence scope. However, there is no authorized measurement of pipe significance either. The influence scope may be measured by hydraulic (e.g. decrease pressure), affected population etc. Although it includes multiple aspects, hydraulic influence is one of the important aspects in assessing a pipe's criticality. The more significant pipe often has greater influence scope to the entire network. The parameter to reflect pipe significance usually answers such questions: if there is some change (e.g. diameter change, pipe close, pipe burst, etc.) in a pipe, what's the effect on the whole system? Pipe's significance is determined by the pipe change's influence or failure consequence. Due to the diversity of failure's definition and consequence, pipe significance assessment is a multi-attribute decision problem. It is simplified to be one-attribute problem which focuses on the hydraulic influence in this research.

Arulraj and Rao (1995) introduced the concept of significance index (SI), which is a criterion that can be applied heuristically to prioritize pipe rehabilitation in the respect of pipe's hydraulic importance. A critical pipe is defined as the pipe most sensitive to the change of Hazen William C value or diameter D. The SI of the j-th pipe is as follows:

$$SI_j = \frac{Q_j L_j}{C_j D_j}$$ (4. 7)

Where Q_j is pipe flow, L_j is pipe length, C_j is Hazen-William coefficient, D_j is pipe diameter. This equation is easy to be understood and can be calculated explicitly without the solution of a linear system of equations.

Pipe index (PI) can also be used to measure the significance of a pipe or the consequence of failure (Vairavamoorthy and Ali, 2005).

$$PI_j = \sum_{i=1}^{n} \frac{\partial H_i}{\partial D_j}$$ (4. 8)

Where, H is the water head, i is the node number, j is the pipe number, and n is the total node number. The indicator reflects that the j-th pipe diameter changes cause the total pressure change. If this partial differentiation value is high, PI illustrates the j-th pipe is important.

Both SI and PI describe the hydraulic influence. Vairavamoorthy and Ali (2005) proved the good correlation between PI and SI (average correlation coefficient R is around 0.80). In addition, the calculation of SI is simpler than that of PI, so SI is preferred as the measurement of pipe's significance in our research.

4.3 Methodology for Pipe Criticality Assessment

Pipe criticality assessment is a two criteria decision making problem which involves pipe condition and pipe significance criteria. Pipe condition refers to the integrity of physical structure. It is mainly affected by the pipes physical characteristics, operation and surrounding conditions. Pipe significance refers to the carrying capacity of the pipe and its importance to flow and pressure distribution within the network.

4.3.1 Introduction of MCDM

Multiple criteria decision making (MCDM) refers to making decision in the presence of multiple, usually conflicting criteria.

Ranking the alternatives by these criteria (attributes) is one of the applications of MCDM. The difficulty of the problem originates from the compromise of the diversified and even conflicted criteria. MCDM problems consist of a finite number of alternatives, explicitly known in the beginning of the solution process. Each alternative is represented by its performance in multiple criteria. A MCDM problem can be concisely expressed in matrix format as:

	C_1	C_2	$\cdots$	C_n
A_1	x_{11}	x_{12}	$\cdots$	x_{1n}
A_2	x_{21}	x_{22}	$\cdots$	x_{2n}
$\cdots$	$\cdots$	$\cdots$	$\cdots$	$\cdots$
A_m	x_{m1}	x_{m2}	$\cdots$	x_{mn}

Where A_1, A_2, . . . , A_m are possible alternatives among which decision makers have to choose, C_1, C_2, . . . ,C_n are criteria with which alternative performance are measured, x_{ij} is the rating of alternative A_i with respect to criterion C_j.   In addition, w_j is the weight of criterion C_j and all weights can be expressed as $W=[w_1, w_2, ..., w_n]$.

The main steps of multi-criteria decision making are the following (Jahanshahloo et $al.$, 2006):

(1) Establishing criteria that relates the system capabilities to its goals;
(2) Generate alterative systems for attaining the goals;
(3) Evaluate the alternatives according to the evaluation criteria;
(4) Applying a normative multi-criteria analysis method; and
(5) Ranking all the alternatives and proposing one alternative as ''optimal'' (preferred).

For step (4), a decision maker should express his/her preferences in terms of the relative importance of criteria. These weights in MCDM provide the opportunity to model the preference structure or criterion importance. The criteria are usually expressed in different units (non-commensurable) and non-dimensional-normalized is often needed. In most cases, it is difficult to determine the values of such weights because decision maker's preference is vague or may not even have an idea about them.

TOPSIS (Technique for Order Preference by Similarity to an Ideal Solution), one of the most classical MCDM methods, is a useful technique in dealing with multi-attribute decision making (MADM) problems. It is proposed by Hwang and Yoon (1981). Based on the criteria and a set of data, a virtual positive-ideal solution/alternative consisting of the best criteria data, and a virtual negative-ideal solution consisting of the worst criteria data are generated respectively. The alternative's rank is determined by the Euclidean distances from the point of the alternative to the point of the positive and to the negative-ideal solution (Opricovic and Tzeng, 2004).The basic principle is that the chosen alternative should have the shortest distance from the ideal solution and the farthest distance from the negative-ideal solution.

Because it is simple and easy to understand, TOPSIS is applied to solve the decision problems in economy and management. Moreover, TOPSIS has solved many real-world problems due to its logical reasoning. For example, Shih *et al.* (2007) illustrated eleven quite different typical applicable areas.

Attributes represent the different dimensions from which the alternatives can be viewed. In the case of pipe criticality assessment, each pipe is an alternative and each criterion is an attribute. TOPSIS is selected as the main methodology for pipe criticality assessment in this study. One reason is that each criterion in TOPSIS should be irrelevant. Pipe condition and pipe significance are independent to each other. Another reason is that pipe criticality assessment is applied for pipe rehabilitation decisions and so that the measurement of pipe criticality should be as simple as possible. Therefore, TOPSIS is a suitable methodology for pipe criticality assessment. Its advantages will be narrated in Section 4.3.3.

4.3.2 Process of Pipe Criticality Assessment by TOPSIS

With TOPSIS method, the criticality indicator should reflect both pipe condition and significance which constitutes the two dimensions of criticality. There are two key parameters in assessing criticality based on TOPSIS. One is the extreme ideal solution, and the other is the weight of pipe condition and pipe significance.

1. Positive Ideal Solution (PIS) and Negative Ideal Solution (NIS)

If these two criteria are depicted in a two-dimension coordinate system, the PIS point represents the ideal pipe which has the best condition (i.e. brand new pipe) and the least significance index simultaneously. In contrast, the NIS point is the pipe with the poorest condition and the highest significance index simultaneously.

2. Weight of Pipe Condition and Pipe Significance

It is difficult to make convictive judgement of which one is important, pipe condition or pipe significance, in pipe criticality assessment. Therefore, the subjective weights assignment is not applied in this research. Only two objective weight assignment methods, coefficient of variation and entropy weighting, are applied. The weights can be considered as the modification coefficients of the two coordinates. They are multiplied with the two values of significance index and condition index after normalization. The addition of weights results in a stretch of each coordinate.

The conventional TOPSIS procedure consists of the following steps (Opricovic and Tzeng, 2004):

(1) Calculate the normalized decision matrix. The normalized value r_{ij} is calculated as:

$$r_{ij} = \frac{x_{ij}}{\sqrt{\sum_{j=1}^{J} x_{ij}^2}} , \quad (i=1,2,3,\ldots n, \quad j=1,2,3,\ldots J) \tag{4.9}$$

Where, i is criteria number and j is alternative number.

In Shih et al. (2007), a few common normalization methods are organized. These are classified as vector, linear, and non-monotonic normalization to fit real-world situations under different circumstances. In our study, vector normalization is adopted to eliminate the units of criterion functions.

(2) Calculate the weighted normalized decision matrix. The weighted normalized value z_{ij} is calculated as:

$$z_{ij} = w_i r_{ij} , \quad (i=1,2,3,\ldots n, \quad j=1,2,3,\ldots J) \tag{4.10}$$

Where w_i is the weight of the i-th attribute or criterion, and $\sum_{i=1}^{n} w_i = 1$,

(3) Determine the positive-ideal solution Z^+ and negative-ideal Z^- solution respectively

$$Z^+ = \{z_1^+, z_2^+, ..., z_n^+\} = \{(\max_j z_{ij} \mid i \in I'), (\min_j z_{ij} \mid i \in I'')\} \qquad (4.\,11)$$

$$Z^- = \{z_1^-, z_2^-, ..., z_n^-\} = \{(\min_j z_{ij} \mid i \in I'), (\max_j z_{ij} \mid i \in I'')\} \qquad (4.\,12)$$

Where I' is associated with maximum-best (e.g., benefit) criteria, and I'' is associated with minimum-best (e.g., cost) criteria.

(4) Calculate the separation measures, using the n-dimensional Euclidean distance. The separation of each alternative from the positive ideal solution (PIS) is given as

$$D_j^+ = \sqrt{\sum_{i=1}^n (z_{ij} - z_i^+)^2}, \qquad (j=1,2,3,...J) \qquad (4.\,13)$$

Similarly, the separation from the negative ideal solution (NIS) is given as

$$D_j^- = \sqrt{\sum_{i=1}^n (z_{ij} - z_i^-)^2}, \qquad (j=1,2,3,...J) \qquad (4.\,14)$$

Some popular measurement of distance for TOPSIS is summarized in Shih (2007) (Shih *et al.* 2007).

(5) Calculate the relative closeness to the ideal solution. The relative closeness of the alternative a_j with respect to Z^+ is defined as

$$C_j = \frac{D_j^-}{D_j^+ + D_j^-}, (j=1,2,3,...J) \qquad (4.\,15)$$

Where, C_j is the relative closeness to the ideal solution.

(6) Rank the preference order.

If C_j is larger, the j-th alternative is closer to the positive-ideal solution. Eq.(4.15) represents the "basic principle" in the TOPSIS method (Chen *et al.* 1992). According to the preference rank order of C_j in the final step, the best (optimal) alternative can now be decided.

4.3.3 Advantage of TOPSIS

TOPSIS is a utility-based method that compares each alternative directly based on the evaluation matrices and weights. According to the simulation comparison, TOSPSIS has the fewest rank reversals among eight common MCDM methods in the category (Shih *et al.,* 2007). Rank reversal depends on the relationship between the new alternative and the old ones under each criterion (García-Cascales and Lamata, 2012).

The four main advantages in applying the TOPSIS method are as follows:

(1) A sound logic;
(2) A metric that measures the best and the worst alternatives simultaneously;
(3) A calculation process that is simple and straightforward; and
(4) Performance measures that can be visualized in a polyhedron, at least for any two dimensions.

TOPSIS has some special strength to solve pipe criticality assessment problems:

(1) The computation process is simple and straightforward, which is suitable for the comparison of large amounts of alternatives (pipes). The pipe criticality assessment is a two-criteria, large amount alternative (pipe) problem. It is a bridge process between pipe condition, pipe significance assessment and the rehabilitation priority decision. Therefore, the computation for an individual pipe is not very complex and the total computation load is not so huge. The advantage of dealing with numerous attributes and complicated hierarchical structure for some of MCDM methods (e.g. AHP) is not necessary in this issue. In contrast, TOPSIS is easy to understand and suitable in this requirement.
(2) Decision makers' preference or subjective influence can be of little effect in TOPSIS if objective weight assignment is applied. In contrast, subjective judgment may have a great effect on numerous MCDM methods (e.g. AHP, PROMETHEE).

4.4 Disadvantage and Modification of TOPSIS

4.4.1 Disadvantage

TOPSIS is based upon the principle that the chosen alternative should have the shortest distance from the positive ideal solution (PIS) and the farthest from the negative ideal solution (NIS). Only if the two conditions are met simultaneously, it can be thought that one alternative is better than the other. Otherwise, although a conclusion can be obtained through Eq. (4.15), the though based on relative distance (i.e., Eq. (4.15)) is different from the principle. The rank by relative Euclidean distance ratio sometimes does not reflect the real rank (Hua and Tan, 2004; Abo-Sinna and Amer, 2005). If the alternative is far away from NIS at the cost of having a longer distance from PIS, there may be two different conclusions.

In order to make the problem clearer, a double-criteria decision making is taken as an example (Figure 4.2). The point A (x_1^-, x_2^-) in Figure 4.2 represents the NIS and the point B (x_1^+, x_2^+) represents the PIS. Other points, C (x_{1C}, x_{2C}), D (x_{1D}, x_{2D}), E (x_{1E}, x_{2E}), F(x_{1F}, x_{2F}) represent four alternatives points. Points C and D are located in the same perpendicular of line AB, and C is farther away from the line AB than D. Point E and F are located in another perpendicular of line AB, and F is farther away from the line AB than E. Point P and Q are the two foot points respectively. According to the traditional TOPSIS computation process, which is introduced in Section 4.3.2, it can be proved that C is better than D, and E is better than F. However, such a conclusion is not reasonable. The distance from C to A (i.e. NIS) is longer than that from D to A, while the distance between C and B (i.e. PIS) is longer than that between D and B. Similarly, the distance from F to A is longer than that from E to A, while the distance between F and B (i.e. PIS) is longer than that between E and B. According to Eq. (4.15), C is better than D, while E is better than F. In contrast, E and D have a similar status, which are close to both points of A and B, but the rank of the two pairs of samples is contradictive.

The main reasons of the contradiction:

(1) The implication of Eq. (4.15) is not exactly same as the principle of TOPSIS. The principle of TOPSIS has such an implication: the best alternative (i.e. the point representing the most critical pipe) is closer to the PIS point and also farther away from

the NIS point than any other alternative. Only if the two conditions are met at the same time, can the rank be judged. On contrast, Eq. (4.15) can also be used for the situation that an alternative is farther away from (closer to) the both points of PIS and NIS. However, this extension is a little different from the original meaning of TOPSIS.

(2) The transformation of multiple criteria to a single criterion is a process of dimensionality reduction, in which some less important information is ignored. In any multi-criterion decision making problem, there is a process that converts multi-criterion to a single and comprehensive criterion if an absolute evaluation conclusion should be drawn. The process of dimensionality reduction is also the process of refining the criteria. In this method, the straight line AB represents the shortest distance from PIS to NIS. The change of distance along this line also represents the most effective part of the change. And the distance change in the direction perpendicular to the line AB does not have a fundamental influence on the judgment. In general, the ignored information has little effect on the judgment.

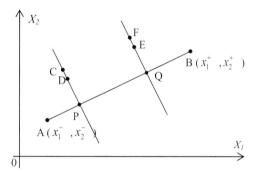

Figure 4.2   Solutions in double-criteria decision making

4.4.2 Modification

For an MCDM problem, a representative and comprehensive criterion is needed so that the original multi-dimensional judgment criteria can be substituted. For multidimensional solution space composed of multiple criteria, it needs to be simplified to a one dimensional space. Only in this way can the criteria be simplified. In this study, a vertical project ion method is applied to reduce dimensions. The basic idea is to project high-dimensional data to low dimensional subspaces, and find out projections that reflect the characteristics of the original high-dimensional data, so as to analyse high-dimensional data. Although the dimension of pipe criticality assessment is a two-dimensional problem, it is necessary to convert the two-dimensional space to one dimension, in order to get a crisp judgement conclusion. In other words, the pipe condition and pipe significance criteria are integrated into a criticality assessment criterion.

In the original multi-dimensional space, PIS and NIS represent the two ideal extreme situations. The line connecting the PIS and NIS points represents an integrated one dimensional criterion. In the original two-dimensional space, the connection between the PIS and the NIS points is the coordinate axis of the new criterion, which is exactly the coordinate axis of the pipe criticality assessment. Therefore, the projection method is to project the points on the two-dimensional space of pipe condition and pipe significance to the axis of pipe criticality assessment. The distance between the projection point and the NIS and PIS points is complementary. If the point in the original two-dimensional space moves along the vertical line of the criticality assessment axis, it does not change the criticality assessment.

This modification can be explained in Figure 4.2. The line section of AB provides the true measurement scale of a solution because the points of A and B represent the NIS and PIS respectively. Every alternative has a projection point on the line of AB. The sum of distance between the projection point to PIS and that to NIS is a fixed value (i.e., the distance from NIS point to PIS point). If the projection point is closer to the PIS, it must be farther away from the NIS. The distance between a solution (a point in the solution space) and line of AB has no effect on the decision because it is a vertical distance on which changes have no impact on the projection point movement on line AB. Actually, the projection point on the line section of AB indicates the effective position which addresses effective distance to the PIS and NIS. In this modification, Euclidean distance in Eq. 4.15 is replaced by the distance between the projection point and the ideal point. Namely, orthogonal projection points on the line AB substitute the original points.

In the example of Figure 4.2, if the vectors of $\overrightarrow{OA}$, $\overrightarrow{OB}$ and $\overrightarrow{OC}$ are $\vec{a}$, $\vec{b}$ and $\vec{c}$ respectively, the vertical projection distance d between A and C are as follows:

$$d = \frac{|(\vec{c}-\vec{a})\cdot(\vec{b}-\vec{a})|}{\|\vec{b}-\vec{a}\|}$$

(4. 16)

Where, • is the dot product, $|\ |$ is absolute value, and $\|\ \|$ is 2-norm.

The projection transformation avoids the argument that is caused by the relative distance. This modification implies the following thought: a vector from NIS to an alternative can be decomposed into two parts, one is perpendicular to the vector of $\overrightarrow{AB}$, another is a projection on vector of $\overrightarrow{AB}$. The former has no substantial impact on the effective distance to the NIS point (or PIS point), which can be ignored. However, the latter has a direct impact on the effective distance, which exactly follows the original TOPSIS principles. The effective distance to the NIS point and that to the PIS point are exactly complementary.

The distance between the projection points to NIS or PIS points, instead of the Euclidean distance in Eq. (4.15), has the following advantages:

(1) The main information of the original multi-dimensional (multi-criteria) is simplified into one-dimensional coordinate, which is also consistent with the principle of TOPSIS.
(2) There will be no such a contradiction as C is more critical than D, while E is more critical than F in the example of Figure 4.2.

Pipe condition and pipe significance are the two criteria (dimensions) for judging criticality. NIS represents the best and most unimportant pipe, and PIS represents the worst and the most significant pipe. In the actual pipe network system, the pipe with the two extreme features is rarely found. For a specific pipe network, the maximum values of pipe condition assessment and pipe significance assessment are the coordinates of PIS in the original assessment space, while the minimum values of those are the coordinates of NIS in the original assessment space.

4.4.3 Weights in TOPSIS

In MCDM, the weights associated with different indicators reflect their relative importance to other indicators.

One category is subjective weights assignments which are generated from experts' experience directly or indirectly, such as Delphi and AHP. The advantage of these is that experts can present or transfer their experience and knowledge to rank the relative importance of each criterion. However, it is greatly affected by individual's knowledge and bias. If there is some wrong experience or bias, they can damage the judgment. In addition, different judgments bring some controversy as well. The subjective judgment about the weights is very uncertain in pipe criticality assessment issue. Almost nobody is able to give a crisp and confident answer to such a question: "How much is pipe condition more important than pipe significance in criticality assessment?" Different persons may give quite different answers or even no answers. Their confidences to their answers vary greatly as well. Such a questionnaire is also difficult to carry out in engineering practice. Therefore, subjective weighting methods are not suitable in this case.

Another category is objective weighting methods which utilize some implicated information in the data to evaluate weights, instead of expert's assessment. For example, entropy weighting, coefficient of variation weighting and principal component analysis are typical objective weighting methods. A common character is that the parameter with greater variance is more important because such a parameter better reflects the difference. If one indicator creates more difference, this indicator has more weights. Because these weights are all generated from objective data, the method is called objective weighting method. The foundation of objective weight method is real life data which are free from subjective judgement. This is also its advantage. On the other side, the disadvantage is that the weights are often contrary to actual problems. Theoretically, the most important criterion is not necessarily the greatest weight, while some unimportant criterion often has great weights. In the case of pipe criticality assessment, what is really needed is the relative rank order of pipe criticality. The parameter with greater variance provides more information to distinguish the difference. Even if the weights might be contrary to the nature of a thing, this is only to distort the distance but not to make the rank reverse.

Subjective weight assignment and objective weight assignment are two typical methods. Subjective weight assignment methods heavily depend on subjective opinions. The bias or the lack of experience may make the weights unreasonable or unfair. Objective weight assignment methods, such as Coefficient of Variation, only rely on the observation data. If the variation of a factor is great, such a factor has a higher weight. The objective weighting assignment methods will be applied to avoid subjective influence in this study.

Two typical objective weighting methods, coefficient of variation and entropy weighting are applied in this research. Although there is no absolute proper weight assignment method, the average weight value from each method is employed as the integrated weights in order to prevent a larger error.

1. Coefficient of Variation Method

The principle of coefficient of variation method is to assign more weight to an indicator with more difference (i.e. coefficient of variation) in the assessment system. It is defined as the ratio of the standard deviation to the mean value. Each indicator's weight is the ratio of its coefficient variation to the sum of all the coefficient variations. Because these indicators can reflect more differences among all the pipes, they are more important and assigned more weight.

Suppose there are m pipes and n criteria. In variation coefficient method, the i-th criterion's variation coefficient is

$$V_i = \frac{\sigma_i}{\bar{x}_i}$$

(4. 17)

Where, σ_i is the i-th criterion's standard deviation, $\bar{x}_i$ is the i-th criterion's average value. The weight of each criterion is:

$$w_i = \frac{V_i}{\sum_{i=1}^{n} V_i}$$

(4. 18)

2. Entropy Weighting Method

The entropy weighting method is based on information theory. It assigns a small weight to a criterion if the values of criterion are very similar across alternatives. The reason is this criterion does not help in differentiating alternatives. In other words, such an attribute should be assigned a very small weight (Xu, 2004). The principle is assigning weights according to the difference of criteria. If some criterion's entropy is small, its variance is great and provides more information. Such a criterion has more influence to the assessment and was assigned a greater weight. In contrast, a criterion will be assigned a small weight.

In entropy weight method, the i-th criterion's entropy H_i is

$$H_i = -k \sum_{j=1}^{m} f_{ij} \ln f_{ij} ,$$

(4. 19)

Where, $f_{ij} = \frac{x_{ij}}{\sum_{j=1}^{m} x_{ij}}$, $k = 1/\ln m$, and assume $f_{ij} \ln f_{ij} = 0$ when $f_{ij}=0$. The entropy weight of the i-th criterion is

$$w_i' = \frac{1 - H_i}{n - \sum_{i=1}^{n} H_i} \qquad (4.20)$$

The average value of these two weighting methods is taken as the ultimate weights for each factor (pipe condition and pipe significance).

$$\overline{w}_i = \alpha w_i + (1 - \alpha) w_i' \qquad (4.21)$$

Where, w_i, w_i' are the weights of the *i-th* criterion for variation coefficient method and entropy weight respectively; α, *(1-α)* are the preference coefficient provided by a decision maker for each method respectively. If there is no preference for each weight assignment method, $\alpha = 0.5$.

There is no absolutely correct method to assign weights to multiple attributes. Neither is the rationality of the weighting method absolute. In the case of pipe criticality assessment, it is not so difficult to assign weights to the two criteria. If the decision maker is very clear which one is the major driving force to pipe rehabilitation, pipe condition or significance, the subjective weighting method can be applied. Otherwise, the subjective bias should not be involved. Whether the decision maker's judgment is clear or not, objective weighting methods are to be applied.

4.5 Summary

This chapter presents a pipe criticality index model that estimates the criticality of individual pipes. The criticality index model is an MCDM one, where the indexes are generated by combining the outputs of two models, a condition assessment model and a hydraulic significance model. The criticality index model is used define the whole-life costs pipe rehabilitation optimisation model presented in the next chapter.

Nominal breakage rate, which is derived from a group of pipes' breakage number prediction, with the emphasis of the individual pipe's historical breakage rate, is the indicator used in this study to describe pipe structural condition. Since a pipe may belong to different pipe groups, its nominal breakage rate can also be derived from these different pipe groups through weighted average. Pipe hydraulic significance is generated by means of a hydraulic analysis of the network, where significance indexes for are combining the flows in the pipe with its physical characteristics.

TOPSIS is selected as the approach to integrate the two components of pipe condition and significance. There are many advantages in using the TOPSIS approach, in particular for the pipe criticality assessment problem. One is that the computation process for an individual pipe's criticality is not so complex and it is easy to deal with numerous water mains in a WDS. Another is that subjectivity is minimized as weights are determined through objective assignments.

The chapter presents a modified TOPSIs approach to overcome the disadvantage in the traditional method, where a pipe's criticality rank by Euclidean distance ratio sometimes is not consistent with the TOPSIS principle. A modified TOPSIS approach, which is based on vertical projection distance, is applied to combine pipe condition and pipe significance to generate pipe criticality index. After normalization and weighting, the projection of a pipe's condition and significance on an ideal standard axis will replace the Euclidean distance in a multi-dimension.

The outcome of the modified TOPSIS method is the generation of criticality indexes (CI) for each of the pipes in the WDS. The ratio of pipe condition and significance after normalization and weighting determine the recommendation of the rehabilitation approach for the pipe, where pipes with high CI are prioritized

Chapter 5   Optimal Rehabilitation Decision Model

5.1 Introduction

Pipe deterioration is an inevitable and continuing process which can be expensive to deal with. Water distribution system's performance maintenance and improvement, depends on planned rehabilitation strategies constrained by budget limitations. Rehabilitation decisions affect the network's performance, both in terms of customer service levels and customer charges. Hence it is imperative the cost-effective decisions are taken when executing a rehabilitation strategy.

In this thesis a whole life cost approach is taken because decisions must consider present conditions and future developments and changes.

The whole life of a WDS usually means the period from the beginning of the WDS service to its end. However, in practice, the beginning time for a whole life analysis, is the present time for an existing WDS and the ending time is a specified point of time in the future.

Generally, optimization design, optimal rehabilitation, or integrated optimization design and rehabilitation are a complex, multi-objective and multi-stage optimization problem. The minimization of costs and the maximization of benefits are often a pair of basic contradictory objectives. Moreover, costs and benefits may be explained and defined differently. How to find the representative and comprehensive objectives among the complex objectives will be discussed in this chapter. In addition, optimal decisions are not made at once, but may be adjusted with development and changes. Therefore, decision-making is carried out in multiple stages from present to future. This chapter will discuss the challenges in developing objectives and decisions that reconcile the needs of the present with the needs of the future.

In the field of optimisation, computational efficiency, convergence and the ability to search for global optimum has received much attention. This chapter will consider the developments new and improved algorithms for the optimization computation of water distribution network rehabilitation. The thesis proposes and improves an optimization algorithm based on NSGAII.

This chapter focuses on WDS network's optimal rehabilitation decision making methodology and process which are based on the concept of whole life costing. The general optimization design and rehabilitation models are reviewed, including the characteristics of optimal rehabilitation decision and the difference between optimal design and rehabilitation. Rehabilitation strategy is a multiple objective decision making process. The diversified objectives are summarized and converted into some representative and measurable indicators for the calculation and decision be feasible. The elements of modelling, including assumptions, general objectives, constraints, decision variables and decision foundations, are addressed in the next section. Because of the decision premise difference, the objectives and constrains in present stage and future stages are discussed separately. Thereafter, the optimization algorithm and its process for present stage decision making is introduced in detail, followed by optimization algorithm for future stages. Next, the combination of present and future stages' decision will be made. In the following section, the possibility of losing some optimal decisions (or solutions) and "no-regret" decision are discussed.

The main content of this chapter can be summarized in the diagram presented below (Figure 5.1).

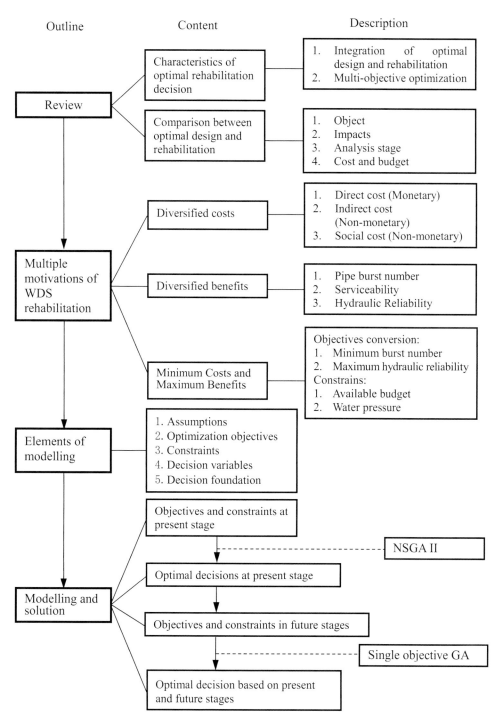

Figure 5.1   Flow chart of optimal rehabilitation decision modelling

5.2 General Optimization Design and Rehabilitation Model Review

Water distribution optimization design and rehabilitation research have been developed with optimization algorithm and practical engineering. The following is a brief review of the research development on this topic.

5.2.1 Foundation of Optimal Rehabilitation Decision

This section starts with the implication of performance indicators of water distribution network, and then introduces the development of the hydraulic calculation method of WDS and optimization algorithms in WDS. All of these are the technical foundation of the WDS optimal rehabilitation decision model.

1. **Hydraulic Computation Theory of Water Distribution Network**

The core problem of steady-state hydraulic computation for all loop water distribution networks is the solution of a large linear and non-linear mixed system of equations. The equations are established according to three classical principles, namely, mass conservation equations, water head loss equations, and energy conservation equations. The common methods used in the calculation of water distribution network are Hardy Cross algorithm, Newton-Raphson algorithm, and Linear Theory algorithm.

The Global Gradient Algorithm (GGA) was proposed by Todini and Pilati (1988). The equations are obtained by the energy equation and the nodal continuity equation. An efficient iterative framework is used to accelerate the convergence of the inverse matrix of the original coefficients. This method was developed by USEPA, and EPANET software was adopted.

Since then, many researchers have proposed methods to accelerate steady state hydraulic calculation. Large scale water supply network topology is usually the combination of loop and branch. The computing efficiency is low if a nonlinear calculation engine is employed for solving linear and nonlinear problems of the whole network. The efficiency of the computation engine will greatly improve if a linear method is applied to the calculation of the flow and pressure of the network (Spiliotis and Tsakiris, 2011). Simpson *et al.* (2014) proposed Forest-Core Partitioning algorithm to speed up the hydraulic calculation of the pipe network.

In the water supply network model, the water demand is assumed to be consumed at both ends of the pipe.   The assumption is different from that of the actual pipeline, where water is distributed along the pipe, but it provides the solvability of the model. GGA was developed to a new algorithm, EGGA (enhancing of GGA) model, considering the errors between the assumption of water distributed at both ends of a pipe and along the pipe (Berardi *et al.,* 2010; Giustolisi, 2010).

2.   Optimization Algorithm Theory in Water Distribution System

Since the 1960s, optimization technology for water supply network has been studied in theory and practice. Water supply network optimization theory and technology has been constantly developed, from branch network to loop network, from consideration of only thr pipe diameter to the integration of a variety of elements (pipe diameter, tank, pump, etc.), from the method of linear programming and nonlinear programming to the random search method, and then to the heuristic search method (Lansey, 2000).

Numerous optimization algorithms, especially evolutionary algorithms, have been developed and applied in solving water distribution optimization problems in the past decades. The general development trend of optimization algorithm is from single objective optimization to multi-objective optimization, from random to heuristic algorithm, from single evolution strategy to the hybrid strategy of optimization.

Some single-objective optimization algorithms have been widely applied to solve water distribution system (WDS) problem as long as time. Single objective optimization is the transformation and simplification of multi-objective optimization of practical WDS problem. These transformations simplify the problem, reduce the objective dimensions and reduce the computation load. However, the transformation and scaling methods often affect the optimal solution significantly (Wu and Simpson, 2002; Wu and Walski, 2004). Moreover, single objective optimization does not reflect the real complex characteristics and relationships in a water distribution system. Rehabilitation decision making of WDS is a typical multiple stage and multi-objective optimization problem.

The heuristic algorithm can be understood as the approximate optimal solutions found at acceptable computation time cost within search space. However, the degree of deviation between the near optimal solutions and the real optimal solutions cannot be expected.

Heuristic algorithm mainly imitates natural algorithm, such as genetic algorithm, ant colony algorithm, and simulated annealing method.

Meta-heuristic algorithm mainly refers to the general type of heuristic algorithm. This kind of algorithm does not depend on the organization structure information of the objectives. Meta-heuristic algorithm includes tabu search algorithm, simulated annealing algorithm, genetic algorithm, and colony optimization algorithm, particle swarm optimization algorithm etc.   Random search techniques are usually used. They can be applied to a very wide range of issues but search efficiency cannot be guaranteed.

Hyper-heuristic is an advanced form of heuristic algorithm, which belongs to a class of mixed optimization strategy. The optimization process of hyper-heuristic algorithm can adjust the process automatically according to the characteristic of the problem (Burke *et al.,* 2003). In other words, different optimization operators (such as crossover and mutation) can be selected according to different optimization problems. The hyper-heuristic mechanism can make one algorithm apply to a wider range of problems and choose a more appropriate heuristic algorithm so as to accelerate the convergence speed of optimization (Burke *et al.,* 2013). Hyper-heuristic algorithm has another kind of hybrid optimization strategy, that is, a variety of optimization methods are coupled together (Grobler *et al.* 2010).

Evolutionary Algorithm (EA) is one of the fastest developed optimization algorithms. It includes Genetic Algorithms (GA), Evolutionary Programming (EP), Genetic Programming (GP) etc. (Nicklow *et al.* 2010). Multi objective evolutionary algorithm uses Darwin's theory of evolution in the optimization process.   The general steps are as follows: (1) randomly generate initial population; (2) evaluate the fitness of individuals; (3) select dominant population using Pareto optimal theory; (4) produce off-springs by using crossover and mutation operators; (5) fitness evaluation after mixing of parents and off-springs; and (6) choose the dominant population to cultivate the next generation. Repeat the 3rd-6th steps until the evolutionary stopping criteria are reached.

In evolutionary algorithms, the optimization operator (selection, crossover and mutation) is the bridge between genetic algorithm and evolutionary strategy (Nicklow *et al.,*2010). The selection operators in GA commonly includes: Tournament Selection, Truncation Selection, Roulette Wheel Selection, and Boltzmann Selection. Binary coded GA often uses uniform crossover, which can induce the algorithm to search the unknown decision space.

Evolutionary algorithms have advantages over multivariate non-linear optimization problems, but they require a long computation time in solving the complex problems of water supply networks. Compared with the numerical algorithm, the evolutionary algorithm cannot guarantee the real optimal solution set because of the limitation of evolutionary algebra. It is generally the approximate solution of the optimal solution set (Zitzler *et al.,*2003).

Mahinthakumar and Sayeed (2005) found the algorithm falling into a "bottleneck" period in multidimensional nonlinear solution space when the searching is near optimal solution. The search efficiency can be improved by using the integrated global searching and local searching methods. Although the hyper-heuristic algorithm has obvious effects on the evolution to the optimal solution region, it is still difficult to solve the complex space optimization problem with multiple objectives. Singh and Minsker (2008) proposed the interactive optimization guided strategy being paid attention to gradually. In the process of optimization, human's subjective experience constantly disturb the algorithm solving process, and make the optimization process approach to the reasonable direction continuously. It is thought that computer optimization technology accompanied by expert experience can solve the problems of large-scale water supply networks, Marchi *et al.* (2014).

Srinivas and Deb (1994) first proposed NSGA, which is classified according to individual's rank. The most distinguishing difference between NSGA and general GA is selection operation. Before selection, the non-inferior solutions will be found in current population. All these solutions form the first non-inferior solution layer which is assigned a larger fitness through assumption. In order to keep diversity, these solutions (individuals) share the assumed fitness. By the same method, the other individuals are classified. The shared fitness in lower level is less than that in the upper level. Such a process is continued until all the individuals are classified. The efficiency of NSGA is the utilization of a non-dominated classification programme which simplifies the multi-objectives into one fitness function. By such an approach, a problem with any objective number can be solved. Later on, Deb *et al.* (2000) proposed NSGA II, a fast elitism non-dominated sorting genetic algorithm for multi-objective optimization.

SPEA2 (Strength Pareto Evolutionary Algorithm 2) is an evolutionary algorithm proposed by Zitzler *et al.* (2001). Unlike NSGA-II, it uses pairwise Pareto to dominate comparisons, and determines the fitness of the solution by the quantity of the dominating solution. In addition,

SPEA2 can only use the penalty factor to deal with the constraints in water supply and drainage network problems.

OMOPSO is representative of the multi-objective particle swarm optimization algorithm (Sierra and Coello Coello 2005). It belongs to the optimization algorithm of Pareto optimal based on the theory of the crowd strategy selecting available solutions and different mutation strategies in different particle swarms. It also uses dominant method to limit population size.

IBEA (Indicator-based evolutionary algorithm) is a kind of optimization algorithm based on evaluation index, which uses multi-objective evaluation index instead of non-dominated sorting and selects the optimal solution set (Van Moffaert *et al.,* 2013). Since the computation is time consuming, the computation load of such optimization methods is very large.

MOEA/D (Multi-Objective Evolutionary Algorithm based on Decomposition) (Li and Zhang, 2009) is a new idea on behalf of the multi-objective optimization method. It uses an integration function to transform multiple objectives into a single objective solution, thus avoiding to solve multiple Chebyshev decomposition simultaneously (Hadka and Reed, 2012). The study of Reed *et al.* (2013) provides the most comprehensive diagnostic assessment of MOEAs for water resources, exploiting more than 100,000 MOEA runs and trillions of design evaluations.

A Multi-Algorithm Genetically Adaptive Method (AMALGAM) (Vrugt *et al,.* 2009) is a meta-heuristic of multi method hybrid strategy. The main contribution of AMALGAM is to recognize a set of combinatorial optimization algorithms that are more robust in solving complex problems. Their hybrid methods include NSGA-II, particle swarm optimization (PSO), differential evolution (DE) and Adaptive Metropolis (AM).

The performance of the optimization algorithm is mainly to examine the convergence and diversity of the solution. The indicators are discussed in detail in the literature review of Zitzler *et al.*(2003).

5.2.2 Characteristics of Optimal Rehabilitation Decision

1.   Multi-objective Optimization

In most of past optimization design research and engineering practice, cost minimization has always been the indispensable objective. In almost all the single objective optimization of water distribution system's design, it is often the only objective. However, cost should not be the only concern of an optimization job. The traditional single objective cost analysis neither takes into account the multiple consequences of a system rehabilitation decision scheme, nor considers the relationship between the water distribution system and the environment (Vilanova *et al,.* 2014).

Driven by cost efficiency, water supply security and WDS performance improvement, more objectives are proposed in the WDS optimization model. Multi-objective optimization has been widely applied in WDS design and rehabilitation decision making (e.g., Tanyimboh and Kalungi, 2008; Giustolisi and Berardi, 2009; Olsson *et al.,* 2009). Thereafter, multiple criteria or objectives decision of water supply network design based on optimization strategy is needed, such as reliability (Farmani *et al.* 2005), robustness (Kapelan *et al.* 2005), and risk (Marzouk and Osama, 2017). In the multi-objective optimization design, the costs are usually taken as one of the required objective functions, and then one or more opposite objectives are selected as the contrary or trade-off of the cost (Wu *et al.,* 2013). Fundamentally, the goal of optimization is to pursue the best cost-effectiveness, rather than a single cost minimization. As there are a variety of indicators to measure costs and performances, the definition and measure of cost-effectiveness has become complex.

The goal in a multi-objective optimization is to find a set of non-inferior solutions which forms the trade-off surface, the Pareto optimal front. For a two-objective optimization, it is a Pareto front. For a three-objective optimization, it is Pareto surface. For an optimization with more than three objectives, it is a Pareto hyper-surface.

Two-objective optimization is more popular than three and even more objective optimization in research and engineering practice, although more objectives optimization is necessary in engineering practice. One reason is that it is the simplest multiple objective optimization. Another reason is that more objectives often make decision makers confused and lost in the diversified objective because it is difficult to make trade-offs among different objectives. Moreover, excessive objectives will result in complex searching for solutions. Khu and Keedwell (2005), developed a 6-objective optimisation model (minimization of network cost and that of five critical individual nodal pressure deficits) and found their approach provided solutions that spanned a spectrum of possibilities, compared with optimisation models that use 2-objectives (minimization of network cost and that of total pressure deficit).

Fu *et al.* (2013) used six objectives in their optimization model: (1) capital cost; (2) operating cost; (3) hydraulic failure; (4) leakage; (5) water age; and (6) fire-fighting capacity. They argue that using several objectives allows complex trade-offs to be considered that would not be possible in a lower-dimensional optimization problem. They used visual analytics to explore the various trade-offs.

Although the method can indeed offer more design choices, decision makers might be confused by the complicated objectives and results. Therefore, two quantified parameters are usually chosen as optimization objectives in most research although the objectives are not the same in different models. Because there are great conflicts and complexity in the requirement of more objectives, less and comprehensive objectives to represent more detailed objectives are needed. Effective optimization algorithm is also necessary to find the solutions in multiple dimensions.

2.   The Integration of Optimal Design and Rehabilitation

Water distribution system performances in design phase are fully considered but the performances after water main deterioration are not always considered by model builders. Because infrastructures start to deteriorate after being laid, the designed working condition is an ideal and temporal status. Therefore, pipe deterioration and rehabilitation should be considered before design. However, due to the complexity and computation capability, the deterioration and rehabilitation process were either over-simplified or not accounted for in design. Usually, these optimization models consider various costs (e.g. the initial construction, rehabilitation and upgrading costs, repairs and pipe failure costs) and the deterioration over time of both the structural integrity and hydraulic capacity of every pipe (Siew *et al.,* 2014). The latest development of computation capability makes large-scale computation easier. More researchers have integrated these two processes in design comprehensively (e.g. Naderi and Pishvaee 2017; Shirzad *et al,.* 2017).

Seifollahi-Aghmiuni *et al.* (2013) used a Monte Carlo based simulation model (MCS) to assess the uncertainties in nodal demand and pipe roughness on long-term performance of the network. They showed that an increase in uncertainty of each variable separately causes a decrease in the deterministically-designed network efficiency.

Creaco *et al.* (2016) sub-divide the whole network construction life into several time phases and compared three different approaches for design of water distribution networks: the single-phase design with demand feedback, the multi-phase design without demand feedback and the multi-phase design with demand feedback. The comparison shows that multi-phase design with demand feedback reflects the difference between actual water demand and designed demand timely. This provides a chance to adjust the design and to make the design more reasonable.

Shirzad *et al.* (2017) introduced an approach, which is named dynamic design, for simultaneous optimization of initial design and rehabilitation scheduling of WDNs during their life cycle. In this approach, pipe diameters in the first year and their rehabilitation/replacement in the next years of the expected life of the network are determined considering the nodal demands growth and increase in pipes' roughness.

3. Different Definitions and Measurements of Objectives

Cost, especially monetary cost, is almost always the optimization objective. Except for monetary cost (direct cost), both indirect and social costs should be accounted for as well. Sometimes, the term of benefit is proposed as a contrast objective to cost (e.g. Vamvakeridou-Lyroudia, Walters *et al.*, 2005).

In the research of Kapelan *et al.* (2006), either maximised overall WDS robustness or minimised total WDS risk is used as the second objective except for rehabilitation cost.

Moreover, reliability is often taken as another optimization objective if it is a multiple objective optimization problem. However, the measurement of reliability is not always the same in literatures (e.g. Tee *et al.* 2014a; Vaabel *et al.*2014; Tanyimboh *et al.* 2016; Karamouz *et al.* 2017). Reliability might be measured by some terms, such as the expected number of customer interruptions per year (Dandy and Engelhardt 2004), network resilience index (Banos *et al.* 2011; Creaco *et al.* 2016), diameter-sensitive flow entropy (Liu *et al.* 2014), surplus power factor (Vaabel *et al.* 2014), and the utility's response time to a pipeline failure (Jin and Piratla 2016). Some researches took network's connectivity, breakage number or outage time as the measurement of reliability. The specified definition or measurement depends on the research goals in different researches. With the different definitions, the measurements of some objectives are not the same.

Dridi *et al.* (2009) proposed three performance indicators, which are structural state, hydraulic performance (pressure deficit), and total cost (defined as the sum of pipe replacement cost and the expected cost of pipe break repairs).

In the research of Alvisi and Franchini (2009), the objectives are to minimize the volumes of water lost and break repair costs. Four objective functions are considered, which are overall risk index, infrastructure's condition, assets' level of service and life cycle cost (Marzouk and Osama 2017).

Thus it can be seen that the measurements of optimization objective are diversified.

4.   Complicated Relationships among the Objectives

Completeness of the optimization objective is necessary, but it is obvious that no model can integrate all of these objectives. Generally, the idea of whole life costing accounts for overall costs minimization and performance improvement maximization, no matter how many objectives there are and how to quantify them. One principle of selecting objectives is that significant objectives must be accounted for and some minor objectives be ignored. Another characteristic among the complicated objectives is that these objectives are not independent of each other so that they can be the greatest representatives of the decision-making objectives. The relationships are not always a straightforward positive or negative correlation. Some objectives are in line with each other instead of competitive relation. For example, hydraulic performance becomes better and break number is reduced after pipe replacement. Some relationships between objectives are non-monotonic. For example, the relationship between break number (rate) and direct costs is non-monotonic. More break numbers lead to high costs owing to frequent emergency repair and maintenance. Very low break numbers also require high monitoring and surveying costs because some intensive monitoring and many pro-active jobs need more investment. A moderate break number is pursued in order to keep a relatively low cost.

However, decision variables and objectives are not equally significant. Fu *et al.* (2012), show how sensitivity analysis can help identify a small number of key decision variables that affect network performance and how these key variables can be used to precondition problem formulations. The use of sensitivity analysis is can therefore improve the computationally efficiency when solving WDS problems.

5.   Chain Decision Serials of Multiple Decision Stages

The whole life cost view requires that the whole decision horizon must be divided into a series of discrete stages for analysis. In practice, the maintenance and rehabilitation action is done every year in its whole service life. Therefore, a dynamic point of view is more realistic.

For example, medium-term (e.g. over a time window of 5 years) and long-term scheduling is used to upgrade the network (Tanyimboh and Kalungi, 2008; Alvisi and Franchini, 2009). If the decision only focuses on present stage, near future or consider a short term view, the rehabilitation might not leave a good development foundation for the future.

The decision in each stage must consider both the situation in present stage and possible consequences in the following stages. The previous renewal decision and its corresponding results become the foundation and premise of the next decision in future stages. Therefore, the decision-making sequence forms a chain structure. The current decision, which will affect all the performances in future stages indirectly, should account for both current and long-term decision consequences. As for the close relation between current and future, the current decision and its consequence is the premise of the next stage. Furthermore, the decision and its consequence in the next stage is the new premise of the following stage. Such a chain relation is extended in the network's whole life. These characteristics lead to complicated decision situations. Under such complicated scenarios, optimization objectives and constraints must be proposed for current stage and future stages.

Multiple-objective optimization leads to a group of near non-inferior solutions under some certain conditions. Each of the non-inferior solutions becomes one of the premises for further analysis for the future. The decision of the next stage must be subjected to which decision is taken in the previous stage, because there is more than one non-inferior decision in a multi-objective optimization problem. Each decision will create multiple decision premises for the following decision stage. Theoretically, there is more than one non-inferior decision corresponding to each present decision in the case of multi-objective optimization problems.

Under such complicated serial relations, the trade-off between benefits and costs is a challenge for decision makers. Such a problem becomes further complicated because of current and multiple future stages. The idea of achieving more benefits at the cost of giving up some previous benefits is widely accepted but the problem includes how to quantify the benefits (and costs) and how to balance them. The monetary costs (and benefits) can be discounted into net present value (NPV) through a discount rate if the discount rate is known. Nevertheless, indirect and social costs, which are still difficult to be quantified, cannot be discounted like direct costs. It is difficult to balance the diversified costs even if at the same stage.

6.   Uncertainty

Pipe deterioration or water demand cannot be predicted accurately because of inherent randomness. The future uncertainties bring too many probable scenarios and increase the complexity of the calculation greatly. There is no absolute and perfect optimal rehabilitation alternative suitable for all possible development scenarios because of uncertainty. Because uncertainty is in the development and the probable different decision in each stage, the optimal decision focus on the whole life is very complicated.

In most research, the uncertainty factors are nodal water demands and pipe roughness (e.g. Seifollahi-Aghmiuni *et al.* 2013; Islam *et al.*2014). If system's deterioration is considered, water main break number is also an important uncertainty factor affecting cost and distribution system performance. A basic approach to deal with uncertainty in the future is to test the distribution system's performance under the possible most critical operation mode and working condition. Some researches describe the uncertainties directly, such as the stochastic model in which water demand and pipe internal roughness are assumed to be independent random variables with some probability density functions. The uncertainty was described as risk or robust in some researches (Cunha and Sousa 2010; Raad, Sinske *et al.* 2010; Naderi and Pishvaee 2017). The uncertainty in the future stages makes the trade-off more complicated. Numerical results show neglecting uncertainty can lead to significant increase in the total cost and amount of unsatisfied demand (Naderi and Pishvaee, 2017). Uncertainties in demand, leakage, and break growth rate have a moderate to significant impact on capital and operation costs (Roshani and Filion, 2014).

The problem is how much risk and what risk is worth taking. With uncertainty, quantifying the cost and benefits become more difficult.

7.   Computation Load

The chain decision serials including all decision stages make the computation and decision complicated. The uncertainties in the future also increase the complexity. In order to simulate the uncertainty in prediction, various development scenarios have to be generated. If optimization is required for each scenario serial, the Pareto solution searching will bring great complexity and huge computation load for the future scenarios.

The above points are the main challenges of water distribution network's optimal design and rehabilitation. A comprehensive model with more challenges simultaneously is needed because this is an important approach making the model more practical. Because stochastic model describing uncertainty often brings huge calculation load, efficiency method dealing with calculation is also needed.

5.2.3 Comparison between Optimal Design and Rehabilitation

Water distribution system's optimal rehabilitation decision is different from optimal design in objects, premise, optimization objectives and constraints. The primary difference is that the object of rehabilitation is an old and existing network while that of design is a brand new network (or an existing network's extension). Almost all the other differences come from this issue. Table 5.1 lists their main differences.

Due to these differences, traditional distribution system's optimization methods cannot be applied directly in whole life costing based rehabilitation decision. Some special characteristics and requirements in rehabilitation decision must be considered:

(1) The existing old network is an inherent constraint and precondition for the decision maker. On one side, it denotes each renewal decision can only improve parts of network components' condition or replace part of a pipe sections in one decision stage. It is impossible to renew the whole network and improve the whole system's performance once and for all. On the other side, this is also a constraint which narrows the range of feasible solutions. In contrast, a new distribution network's design does not have such a constraint.

(2) Pipe failure and its impact are seldom considered in conventional design but it is an important deterioration sign and indicator. Therefore, failure number minimization is often taken as a rehabilitation objective.

(3) Cost saving or cost minimization might not be the main optimization objective in rehabilitation decision. For an asset manager, the minimum performance requirements (e.g. pressure) and approximate budget limit for rehabilitation are usually set before making a decision. In such a case, budget limit is one of the constraints rather than optimization objective. Engineers are required to maximize the general benefits or various network performances by taking full advantage of budget. Therefore, cost saving is compromised in such combination of objective and constraints. If funding is so insufficient that the performance cannot meet the requirement, budget should be increased or performance requirement should be lower. In contrast, budget is not the constraint

before water distribution system (WDS) design because the total investment is unknown before design. Therefore, cost minimization is usually the optimization objective in design.

(4) System renewal has to be carried out stage by stage instead of once and for all. Because annual budget is limited and the deterioration is a long-term and gradual process, rehabilitation actions have to be done every year. In addition, the uncertainty also requires the decision to be adaptable to different future scenarios. The traditional design seldom considers the gradual deterioration and uncertainty.

(5) Water distribution network expansion (e.g., the service area is extended, and nodes and links are added) is not considered directly in this research. As an alternative, the extension part of a water distribution network can be simplified as node demand and pressure requirement increasing in some edge nodes. Meanwhile, the original edge nodes' pressure requirements also increase in most cases.

Table 5.1   Comparison between rehabilitation decision and design

No.		Rehabilitation decision	Design
1	**Object**	1. An existing old network	1. A brand new network or an existing network's extension
2	**Impacts**	2. Hydraulic performance	2. Hydraulic performance
		3. Pipe deterioration and failure	3. Pipe deterioration and failure
		4. Water quality risk (not always included)	(not always included)
			4. Water quality risk (seldom included)
3	**Analysis stage**	5. Multiple and chain stages/scenarios	5. Single or multiple stage /scenarios
4	**Cost and budget**	6. Cost minimization is optimization objective or constraint due to budget	6. Cost minimization is often required
		7. Budget is set before decision making	7. Budget is not set before design or plan

These characteristics make the optimization of WDS rehabilitation decision to be different from that of design. The objectives and constraints are complicated than those in conventional optimization design.

5.3 Multiple Motivations of WDS Rehabilitation

There are a variety of motivations for pipe replacement, such as to reduce the volume of lost water in their network by repairing or avoiding main breaks, to alleviate water quality concerns, to improve overall area delivery, or even to take the whole-of-life energy considerations into account (Prosser, *et al.* 2015). The general purpose of water distribution renewal is to gradually restore and improve the performance of the system under an acceptable cost. The direct motivations of WDS rehabilitation are usually decreasing interruption to customers due to higher breakage rates and improving hydraulic capacity due to increased roughness of pipes. Water quality improvement might be an indirect outcome as well. These multiple motivations can be summarised into two (often contradicting) objectives - cost minimizing and benefit maximizing.

5.3.1 Cost

The general concept of cost can be classified into three categories (direct, indirect and social costs).

Direct Costs – refers to repair and renewal costs, cost of water losses, cost of damage to the infrastructure and the surroundings etc. Direct costs are often dependent on the severity and location of the failure.

Indirect Cost – refers to the cost associated with supply interruption and lower levels of water service (economic losses suffered by industry and businesses).

Social Costs – refers to costs associated with water quality degradation, reduced public trust in the quality of the water service, cost of disruptions associated with major bursts and repairs, cost of interruptions to sensitive/special facilities (hospitals, schools, etc.).

The difficulties to quantify these different costs in practice are summarized below (Rajani and Kleiner, 2002):

(1) Direct costs are relatively easy to monetize, but indirect and social costs are more difficult to describe and quantify.

(2) The failure effect has obvious randomness, because no two failures have the exact same consequence.

(3) Failures in large mains are relatively rare, whereas the failure frequency in small mains is usually higher and hence the accumulated data is more abundant to support damage and cost estimation.

(4) The consequence of hydraulic failures is rarely assessed, except when fire liability is concerned.

(5) The consequence of water quality failure is more difficult to assess because it includes the customer's unpleasant feeling, direct and potential health damage etc. It is paid more attention to but rigorous assessments are needed.

All the difficulties can come down to two points:

(1) Some data are not available. Most of the items in these three cost categories (direct, indirect and social costs) depend on the location and outage time of the failure which vary greatly in practical engineering. The raw data and cost functions are usually not ready as well.

(2) The quantification of indirect and social cost is difficult. Almost all direct cost components can be addressed by monetary terms. However, the items in other two cost categories usually cannot be measured by or converted into monetary terms. This causes difficulty of quantifying the overall failure costs.

For these reasons, the rehabilitation objectives must cover these three cost categories. Some conversion and substitution of the objectives are necessary. Usually, the direct cost is contradicted to the indirect and social costs.

In the view of quantification, these three costs can be classified into monetary cost and non-monetary cost.

1. Monetary Cost

The term "cost", usually refers to "direct cost" or "monetary cost" in most cases, is often concerned in WDS optimization. In rehabilitation decision, monetary cost is often subject to budget. In some cases, the planned budget is determined by senior decision makers before actual rehabilitation strategy making. Even if there is not a clear budget limit, the analyser still can assume some different budget thresholds and provide corresponding optimal decisions under different budget constraints. For such a reason, the direct cost is treated as a constraint in the optimization in this research.

The annual budget limit is usually assumed to be crisp and known before decision making. In other words, to minimize economic costs is not necessarily one of the required optimization objectives. In contrast, the pursuit of a better cost-benefit ratio in the range of economic affordability is often a more important goal for a modern water supply enterprise. What the decision maker's concern is how much the performance could improve with the available budget. In other words, the objective is making full utilization with the limited budget. If the performance cannot satisfy the requirement after optimization with a certain budget, either budget limit or performance requirement should be changed. In such an approach, direct cost is converted from optimization objective into a constraint.

2. Non-monetary Cost

As the contrast category of direct cost, both indirect and social costs can be aggregated into one category, which is non-monetary cost. WDS performance improvement is usually consistent with indirect and social cost decrease. The term of performance is also a comprehensive concept including various aspects. An important performance indicator about non-monetary cost in rehabilitation decision is pipe failure. Less pipe failure number leads to less indirect and social costs, such as water supply outage time and less water pressure decrease. Non-monetary costs include diversified aspects that they are difficult to describe and assess.

5.3.2 Benefit

The general benefit can be summarised as the performance improvement of the WDS. The benefit of WDS rehabilitation can be classified into three categories (pipe burst number, serviceability and hydraulic reliability).

1. Pipe Burst Number

Pipe burst number reduction has always been a key issue for WDS rehabilitation. Pipe deterioration often results in structural failure, hydraulic failure and water quality failure. Structural failure is the main type and often dominated the other two. The stress of rehabilitation is pipe structural failure which is measured by burst number or breakage rate of a distribution system. As the representative of structural failure, burst or breakage has close relevance to most of the cost and benefit components. Then, the relationship between the sub-objectives and the burst number could be investigated.

2. Serviceability

The serviceability includes two aspects, one is hydraulic performance or hydraulic capacity, and the other is the ability of continuous water supply. The hydraulic performance mainly refers to water delivery capacity, which is affected by increase of water demand and pipe internal corrosion and tuberculation. The latter results in roughness increase and diameter decrease, that ultimately leads to greater resistance to flow through the pipes and pressure distribution problems. In terms of continuity of water supply, some of the surrogate indicators have been used, such as the number of a day's outage, the volume of undelivered water demand, the number of customers interrupted, the number of failures and the duration in the failure state. Index-based measures have included the ratio of the served demand to the total demand, and the complement of the ratio of the undelivered demand over the total demand.

3. Hydraulic Reliability

The implication of reliability may have different aspects, such as the connectivity between the pipes or network topology, hydraulic reliability and residual energy in the nodes.

Reliability in network topology focuses on the connectivity between the pipes and guarantees that the water can be transported from water source to any point in the network through at least one path (Torii and Lopez, 2012; Gheisi and Naser, 2013).

Hydraulic reliability is to meet a satisfied hydraulic service requirement at some probability under change of condition (Yannopoulos and Spiliotis, 2013; Liserra et al., 2014). The reliability of WDS can be expressed in several ways including: probability that it is operational (reliability); the percentage of time that it is operational (availability); or in terms of surrogate measures that reflect the operational requirements of the system (serviceability).

Liu et al. (2014) carried out the correlation analysis between entropy and traditional methods of reliability evaluation, and pointed out that entropy could be used as an alternative evaluation index of reliability.

The term resilience is also used to quantify the hydraulic reliability and the availability of water during pipe failures (Todini, 2000). Similarly, the residual energy in the network is also used to evaluate the reliability of the water supply network (Vaabel et al. 2014), such as the minimum surplus head available in a network (Prasad and Park, 2004). Herrera et al. (2016) addressed a graph-theoretic approach for the assessment of resilience for large scale water distribution networks. If the terms reliability and resilience are further compared, it can be found that reliability pays more attention to the qualified rate of the service of the water distribution network, while resilience implies the capability of the system to resist external interference. Although they have different implications, the two concepts have high consistency.

Water distribution network reliability is a hybrid measure affected by the network topology (redundancy) and its hydraulic capacity. The multitude of reliability definitions naturally makes it more difficult to define a reliability failure.

Raad et al. (2010) used four reliability indirect indicators, including the resilience index proposed by Todini and improved resilience index, to analyse their possibility to replace reliability evaluation. Baños et al. (2011) proposed two performance indexes, which are average failure probability and the failure condition of minimum water requirement, and study the correlation between the two indexes and resilience. Atkinson et al. (2014) made optimal design of water supply networks with resilience index, entropy and minimum surplus water head. In this research (Atkinson et al. 2014), the increase of resilience and minimum surplus water head can improve the network reliability, while the entropy is more suitable to measure the improvement of the mechanical reliability of the network. Creaco et al. (2016) combined the Todini resilience index with the diameter uniformity index in a loop network, and proposed a new reliability indirect evaluation method.

5.3.3 Objectives Conversion

The diversified objectives usually make optimization models complicated and confuse the decision maker. Therefore, reducing objective number is a necessary and possible approach.

The selected objectives should be representative because the objective selection is the guidance of decision making. One alternative is to search for groups of objectives with the same variation tendency in case of pipe deterioration. A typical performance indicator in each group is selected to represent the others in the same group. These representative and comprehensive indicators are usually the refined optimization objectives. Another feasible idea is to convert some objectives into constraints if the requirements and limits are clear. Both of them are applied in this study.

The selected objectives should also be quantifiable or be replaced by other similar and quantifiable indicators. Among the objectives of cost, direct cost is relatively easy to be quantified by monetary term, but indirect and social costs are not easy to be quantified. Among the objectives of benefit, burst number is a clear indicator but the measurement of other two are not authorised.

Not only the objectives contain a number of sub-objectives, the measurements for the same sub-objective are also different. If all of the sub-objectives are considered simultaneously in optimization decision, it will make decision objectives too complicated, not only for calculation but also final decision making. However, these sub-objectives are not independent. There could be one or a few indexes representing all or most of the other sub-objectives.

Most of the objectives are related with burst number because its decreasing is a main motivation of rehabilitation. If an objective is independent of burst number, this objective can't be replaced by it.

1. Total Cost Minimizing

Generally, economic saving usually contradicts with the other performance improvements. For example, total expenditure and the prevention of future asset failures (serviceability) are taken as two trade-off objectives to identify the optimized investment policy according to the decision maker's priorities (Ward et al., 2017). Direct cost is almost the synonym of economic cost. The indirect and social costs are closely relevant to burst number, serviceability and hydraulic reliability performance. Therefore, it is necessary and possible to sort out the costs and benefits so that fewer indexes represent more objectives.

The relationship between burst number and direct cost is non-monotonic. The moderate burst number is corresponding to an economic burst rate. For an existing WDS, the direct cost mainly consists of two parts: (1) water loss and other related costs (e.g., repair cost) due to leakage and burst; and (2) costs of water main's monitoring and maintenance. If a network is poorly protected and maintained, the first part is high or the second part is low, and the burst rate is high. On contrast, if a network is well protected and maintained, the first part is low or the second part is high, and the burst rate is low. Since the total direct costs mainly consists of these two parts which have contradicting variation tendencies with burst rate (or burst number), the minimum direct cost must correspond to a moderate burst number. However, the exact value of the moderate burst number is unknown. The relationship between burst number and indirect cost is non-monotonic. Both indirect cost and social cost are positively correlated with burst number.

2.   Total Benefit Maximizing

Pipe burst number deceasing is a direct objective. Moreover, it had close relationship with serviceability. The serviceability of the system can improve by pipe replacement or relining, either in hydraulic capacity or continuity of water supply. Pipe relining or replacement can directly change the hydraulic capacity of these pipes. Meanwhile, pipe replacement can reduce the pipe breakage number, thus reducing the impact on the continuous water supply.

Hydraulic reliability is about dealing with uncertainty which is no relation to burst number. The uncertainty here mainly refers to the uncertainty of water consumption. Burst number or burst rate is a surrogate of distribution system's structural reliability. On the other hand, minimum surplus head available in a network or resilience index are mainly used to describe the capability of dealing with the uncertainty or accident in the system. They are another kind of reliability indicator which is no relation to burst number.

Based on the above analysis, the relationships between the primary objectives, sub-objectives and burst number are as summarized in Table 5.2. The two primary objectives are subdivided into some sub-objectives respectively. Since pipe burst number reduction is a key issue, the relations between the sub-objectives and burst number are addressed. Whether each sub-objective can be monetized, and whether there are constraints, is also addressed in Table 5.2.

Table 5.2   Objective decomposition

Primary objectives	Sub-objectives	Relation to burst number	Monetary conversion	Constraint
Min total cost	Min direct cost	Moderate Burst number	Yes	Yes
	Min indirect cost	Min Burst number	No	No
	Min social cost	Min Burst number	No	No
Max total benefit	Min burst number	Min Burst number	No	Yes
	Max serviceability	Min Burst number	No	No
	Max hydraulic reliability	No relation	No	No

Through Table 5.2, it can be found that:

(1) The sub-objective of minimum direct cost can be converted into a constraint. Direct cost is not a proper objective in this case though direct cost (the monetary cost) is a usual optimization objective. The reason is that it requires a moderate burst number, which is not a clear or crisp optimization objective. However, this sub-objective can be converted into a constraint. Because the direct cost is often subject to available budget and the rehabilitation budget limit is determined before decision making in some cases, the minimum direct cost objective is not as important as it is in optimal design. Namely, it can be explained that making full utilization of budget is the real goal instead of cost saving.

(2) Of the six sub-objectives, four can be converted to "Min burst number". Therefore, "Min burst number" can be used as a representative and comprehensive objective after conversion.

(3) The sub-objective of "Max hydraulic reliability (dealing with uncertainty)" has no relation to burst number. Therefore, it should be kept as an optimization objective independent to burst number.

In summary, all of the costs and benefits can be converted into three surrogates: (1) direct cost; (2) burst number; and (3) hydraulic reliability. Therefore, the optimization objectives are summarised: (1) minimum burst number; and (2) maximum hydraulic reliability (dealing with uncertainty). Meanwhile, these objectives are subject to two constraints: (1) available budget; and (2) required minimum water pressure or maximum allowed insufficient pressure.

5.4 Elements of Modelling

5.4.1 Assumptions

In order to make the research focus on the key problem, reasonable simplification must be made and two important assumptions and prerequisites are needed:

(1) Distribution network's topology structure remains unchanged. If the network is extended, the design of the new sections of the network is beyond this study. However, the effects due to the network extension could be simulated through water demand and pressure requirement changes in the nodes connecting old and new network. The new network's water demand and pressure requirement can be passed on to these connection nodes.

(2) Gravity water supply. With such an assumption, the decision making is not affected by the complicated energy consumption. The primary reason of proposing such an assumption is to remove the influence of energy consuming and simplify the calculation. There are two foundations for such an assumption: (1) water source head is usually stable, whether it is gravity or pressure water supply; and (2) water demand is determined by water consumption and has little relevance with rehabilitation. For the first foundation, if water source pressure is allowed to vary in a small range, the stable water sources pressure can be realized through switching pumping combinations in the case of pressure water supply. The pressure is naturally stable if it is gravity water supply. Since the water pressure and demand are almost stable no matter whether the water supply is gravity or pressure, the energy consumption is also stable. For such a reason, energy consumption is almost the same and its influence can be ignored. For example, a gravity water supply network is taken as a design and rehabilitation case study (Jayaram and Srinivasan, 2008).

(3) A particular pipe material is usually corresponding to a particular diameter in case of pipe replacement. Although material may be changed with replacement, a particular material is usually preferred for a particular diameter during a certain period. For example, pipes of 300 mm or above in diameter nowadays are usually ductile iron pipe in water distribution systems.

5.4.2 Optimization Objectives

The general optimization objective is to minimize the rehabilitation costs and maximize the benefits. These objectives are converted into minimum burst number and maximum hydraulic reliability with some constraints. They are independent to each other. The detailed measurements of the two objectives are addressed in the below.

1. Burst Number

In order to calculate direct cost of pipe breaks, a crisp predicted break (burst) number is necessary to be known. A deterministic model for breaks number prediction is suggested as Eq. (3.11) in Chapter 3. It should be noted that Eq. (3.11) is only for the total break number of the pipe group. Since the pipes are usually grouped according to diameter, the total predicted burst number is the sum of each group with same diameter. Multiple non-linear regressions can be used for coefficient values estimation if all of the data is available. Hypothesis testing can be used to test the formula's feasibility.

For the future stages, some parameters in Eq. (3.11) will be updated because some rehabilitation action is taken. For example, the pipe group's equivalent age, freezing index and historical breakage record will be changed with one pipe replacement. If this pipe's diameter is also changed, such a pipe will belong to another different diameter pipe group and the equivalent length will be changed as well. Therefore, the burst number in future stage is closely relevant to the rehabilitation action in each stage. Meanwhile, the coefficients in the formula are derived from current available data. These coefficient values might be changed with pipe deterioration development and new historical data when the future becomes history. However, these changes cannot be predicted until the real new data are obtained.

2. Modified Resilience Index (MIr)

Todini (2000) proposed Resilience Index (I_r) to quantify the hydraulic reliability of a water distribution system.

$$I_r = \frac{\sum_{i=1}^{N_n} q_i^* (h_i - h_i^*)}{\sum_{k=1}^{N_r} Q_k H_k - \sum_{i=1}^{N_n} q_i^* h_i^*} \tag{5.1}$$

Where, q_i^* and h_i^* are water demand and minimum required head respectively at the node i, h_i is actual head at the $i-th$ node, i is node ID, N_n is the number of nodes, Q_k and H_k

are the discharge and the head relevant to each reservoir (water source) k respectively, while N_r is the number of reservoirs.

The essence of Resilience Index is that the greater the water pressure (energy) of each water node is, the stronger the capability to deal with some abnormal conditions, and the higher the reliability of the parameters. In order to facilitate comparison, this part of the surplus energy is expressed in the way of relative value.

Jayaram and Srinivasan (2008) revised it and presented Modified Resilience Index (MI_r) to describe the capability to handle uncertainty in a water distribution network. MI_r is proved to be better than Resilience Index (I_r) in the case of multiple water sources. For such a reason, Modified Resilience Index (MI_r) will be proposed as the measurement of hydraulic reliability in this research. The index is defined as the amount of surplus power available at the demand nodes as a percentage of the sum of the minimum required power at the demand nodes:

$$MI_r = \frac{\sum_{i=1}^{N_n} q_i^*(h_i - h_i^*)}{\sum_{i=1}^{N_n} q_i^* h_i^*} \qquad (5.2)$$

Where, q_i^* and h_i^* are water demand and minimum required head respectively at the node i, h_i is actual head at node i, N_n is the number of nodes. Compared with I_r, the change in MIr is in the measure baseline of the relative value, and is not in the nature of the Ir.

In current rehabilitation decision stage, node pressure can be obtained through hydraulic calculation. Then the MIr can be derived. For future stages, the MIr can be obtained in the similar method. It should be noted that the water demand and pipe roughness are uncertain in prediction and the most probable values of them will be chosen as representatives.

5.4.3 Constraints

All the constraints can be classified into two categories, rigid constraints and flexible constraints. Rigid constraints must be strictly followed, while flexible constraints might be violated to some degree in some cases but the fitness of objectives will be discounted.

1. Rigid Constraints

(1) Conservation of mass. The sum of all ingoing and outgoing flows in each node equals zero.

(2) Conservation of energy. The water head difference between two nodes equals the water head loss along the route connecting the two nodes. The sum of all head-losses along pipes that compose a complete loop equals zero ($\sum \Delta H_i = 0$)

(3) Diameter. The pipe diameter is only available among the standard diameters.

2. Flexible Constraints

(1) Budget Limit (Economic Constraint)

Direct costs, including repair expenditure, breakage damage compensation and rehabilitation expenditure, are subject to available budget:

$$Cost^{dir} < Budget$$

Because decision maker or budget holder does not really understand the exact relationship between the network performance and expenditure before optimization analysis, how much budget is necessary is not made from the consideration of performance but affordability, which is usually flexible to some degree. Hence, such a constraint might be violated to some degree.

It is possible that the performance improves greatly at the cost of a little expenditure increasing. In such a case, it is worth to spend more money, even if the direct cost exceeds the pre-set budget limit. Economic constraints and the penalty of violation are determined by decision maker. The violation degree of the economic performance constraint is easy to be measured if it is violated only if the direct cost and budget is known.

(2) Hydraulic Performance Constraint

It is required that pressure on each node must be greater than the minimum pressure requirement:

$$H_i > H^*$$

Where, H_i is the pressure at node i, H^* is the minimum pressure requirement in each node. Only if there is a node's pressure less than the minimum pressure, this constraint is violated. The minimum pressure requirement might be a certain constraint in a model but a flexible one in most practical cases. Such a crisp constraint is usually proposed according to some service standard instead of some strict natural principles. If violation of the constraint occurs, the penalty of violation will be applied. An immediate thought is that the penalty is in direct proportion to the node number of the insufficient pressure and the degree of insufficient pressure. Therefore, the insufficient pressure's influence can be measured by Insufficient Pressure Index (H_{in}):

$$H_{in} = \frac{\sum_{j \in N_{in}} q_j^* \cdot (h_j^* - h_j)}{\sum_{i \in N} q_i^* \cdot h_i^*} \qquad (5.3)$$

Where, q_i^* and h_i^* are water demand and minimum required head respectively at node i, h_j is actual head at node j, N_{in} is the ID set of the nodes with insufficient pressure, N is the ID set of all the water demand nodes, q_j is water demand at node j. If N_{in} is null, all the nodes meet the minimum pressure requirement and $H_{in} = 0$. Otherwise, $H_{in} > 0$. The index can be regarded as the ratio of insufficient energy to the total energy requirement. If the water demand is thought to be the surrogate of consumer number or influence scope, and the insufficient pressure is thought to be the influence degree on some nodes, the product can be thought as the comprehensive influence of them.

This index denotes general pressure insufficient degree under a specified condition. It is relevant to hydraulic reliability but they are different. High reliability does not guarantee the fair water pressure distribution at all of the nodes. Some insufficient pressure node could be hidden (or offset) by some high pressure. For example, in the definition of Resilience Index (I_r) proposed by Todini (2000), some high pressure in some nodes may cover up the fact of some insufficient pressure on some other nodes if only reliability is considered. Therefore, Insufficient Pressure Index (H_{in}) is taken as a flexible constraint which can be violated to some degree and penalty is attached in such a case.

5.4.4 Decision Variables

Pipe replacement and pipe relining are chosen as rehabilitation approaches because they are representative and most widely used.

1. Replacement

This is a complete and the most expensive rehabilitation action which impacts all of the pipe's features. With such an action, pipe age, diameter, historical breakage record and even pipe material all might be changed. Moreover, pipe break tendency and roughness growth tendency are also affected.

2. Pipe Relining

Only non-structural relining is considered in this study, which reduces the roughness of internal wall and makes it smoother. It does not change the structural condition of the pipe (e.g. pipe break tendency). Except the initial roughness of the pipe after relining, the pipe diameter, age and the deterioration trend will not change. This method has been part of a demonstration program for pipe rehabilitation to evaluate pipe rehabilitation technologies developed by USEPA (Selvakumar and Matthews, 2017).

3. No Action

Leave pipe deteriorate as before and no action is taken.

In this model, there must be and only one of the three actions can be taken for a pipe during a specified period of time.

5.4.5 Decision foundation

1. Pipe Age

Pipe age is usually available in most water utilities databases. It is mainly used in pipe burst number prediction and roughness prediction. Only with pipe replacement, the pipe age in the same position can be changed as well. The function between burst rate prediction and pipe age can be depicted through Eq. (3.12) .

2. Pipe Roughness

For the pipes with and without relining, pipe roughness will grow with age with a little different tendency. In this study, it is assumed that roughness (e) of a pipe increases linearly with age t:

$$e = e_0 + at \qquad (5.4)$$

Where, e_0 is an initial pipe roughness, and a is roughness growth rate. In the absence of sufficient observed data, the pipe wall roughness is assumed to be 0.18mm (as suggested by Sharp and Walski (1988)). The relationship between the Hazen-Williams coefficient and the internal roughness of the pipe is as follows (Sharp and Walski 1988):

$$C = 18.0 - 37.2 \times \log(\frac{e}{D}) \qquad (5.5)$$

Where, D is the pipe diameter. Resistance to water-flow can be obtained through Eq.(5.5). Although pipe relining does not help to reinforce the pipe structure, its internal roughness will be changed and the growth tendency will also be changed. After relining, the pipe age is not affected, which means the burst tendency will not change. If a pipe is replaced or relined, the age for roughness calculation will be treated as a new pipe. The age for burst rate calculation is not changed if relining is applied.

3. Direct Costs

The direct cost of a WDS includes the expenditure and the economic loss due to pipe breaks.

$$Cost^{dir} = Cost^{re} + Cost^{br} \qquad (5.6)$$

Where, $Cost^{re}$ is replacement or relining expenditure, $Cost^{br}$ is economic loss due to pipe breakage. In one renewal decision, a pipe can only be replaced or relined. In the case of replacement, the expenditure can be expressed by a power function:

$$Cost_i^{re} = a_1 \cdot D_i^{a_2} \cdot L_i \qquad (5.7)$$

In the case of relining, the expenditure can be expressed by an exponential function:

$$Cost_i^{re} = a_3 \cdot \exp(a_4 \cdot D_i) \cdot L_i \qquad (5.8)$$

Where, i is the rehabilitated pipe's ID, a_1, a_2, a_3 and a_4 are coefficients, D_i is the $i-th$ pipe's diameter after rehabilitation, and L_i is the pipe length. The coefficients of a_1, a_2, a_3 and a_4 vary according to the rehabilitation method, relining or replacement.

The economic loss due to pipe breakage mainly includes repair expenditure, water loss and damage compensation. All of these items are considered comprehensively as a function of diameter and corresponding breakage number of the network.

$$Cost^{br} = \sum_{k=1}^{K} f(D_k) \cdot BR_k^{es} = \sum_{k=1}^{K} (b_2 \cdot D_k^2 + b_1 \cdot D_k + b_0) \cdot BR_k^{es} \qquad (5.9)$$

Where, D_k is the k-th diameter, $f(D_k)$ is the unit cost of pipe breakage for the diameter D_k, BR_k^{es} is the estimated breakage of the k-th diameter, K is the total number of diameter categories, b_0, b_1 and b_2 are coefficients respectively which can be estimated through curve fitting.

4.   The Time Span of Whole Life

In existing literature, some research (e.g., Jayaram and Srinivasan, 2008) takes 30 years as the life cycle. Engelhardt *et al.* (2003) proposed a period of 50 years. Karamouz *et al.* (2017) proposed a threshold based on the ratio of system revenue and cost to determine the life cycle span of the system. According to the case study, it can be concluded that the efficient lifetime of a WDN can be much shorter than that of its design (the efficient life span is 14 to 16 years, and the design life span is 20 years).

There are some extreme cases of long life pipes in practice, but the span of whole life should not take these extreme cases as a reference. For example, some of the water distribution pipes in London are more than 100 years old. The good pipe manufacture quality is the inherent reason. The stable service environment, well maintenance and careful protection that make the pipes free from damage are the external reasons. Even though pipe aging and deterioration still occurs on these pipes. If the observation time window is long enough, pipe burst still can be found on them.

The specified time span of the whole life is determined by the research objects and purpose. Some key components in the system (e.g. backbone mains) need a longer time view and some common water mains only need a relatively shorter view. In a master development plan, the time span of whole life is longer. While in the shorter-term plan, the time span of whole life is shorter.

There is no authorized definition about how long the whole life is. If the time span is too short, there is little chance to reflect long-term effects. Generally, whole life time span should be long enough to fully reflect the consequences of decision-making. On the other hand, if the time span is too long, there is much uncertainty (e.g. breaks and roughness prediction) in far future. Moreover, the possible performance influence in far future has little impact on present decision making and the current action has little impact on the performance in far future. If the life span in a model is too long and there are many decision stages, both computation load and the impact of uncertainty will increase, the reliability of future analysis will be reduced as well. The life span in a model should be a moderate value.

Time step of each stage mainly relies on different model and decision maker. Engelhardt *et al.* (2003) applied 5-year time steps in line with the pricing period utilized by OfWat (the Water Services Regulation Authority in England and Wales). In the research of Dandy and Engelhardt (2006), it is assumed that the pipes can be replaced at any pre-specified time step over a defined planning horizon.

5.5 Objectives and Constraints in Different Stages

The objectives and constraints in different decision stages are also different. The present rehabilitation decisions must immediately bring good performance to the deteriorated network at least in present stage. Meanwhile, it should provide some capability to adapt for future uncertainties as well. Therefore, the present rehabilitation decision is thought to be important than those in the future. In the whole life costing view, current decision's influence should be considered not only in current stage but also in future stages. Although the optimal decision is wanted in many stages, the premise and objectives are not the same for current and future stages.

The rehabilitation decision making is based on multiple predicted stages, but the action is taken step by step. The rehabilitation action in present stage is to be carried out immediately.

However, the assumed rehabilitation actions in the future stages are possible response plans that are not necessarily put into practice, because the new decision in the future will be made in the same way according to the actual development. The future action is to test present action's impact in the future and it is the chain link in the decision chain serial. In each development stage, the decision view is always based on the long-term but the rehabilitation action is taken only at that stage of time. The real option in the future is not decided in present stage. Real future decision will be made in the future because the uncertainty will be reduced overtime.

In order to reduce the computation load, only some represent scenarios are taken into account. Simplification is employed to deal with the uncertainty. If a typical scenario really happens in the future, a new optimal decision will be made accordingly. Any present decision must be near-optimal at least in the current view. Therefore, there are two comprehensive fitness functions (indexes) as optimization objectives in present stage. Future optimal decision's generation process is simplified and these decision's results is the bridge between previous and following stage. The main function of them is to demonstrate the most probable optimal decision and their consequence so that each current decision's long-term influence can be involved in analysis.

The optimization decision of future stages is mainly to test present decisions' flexibility and adaptability, instead of making some definite decision for the future. Namely, the optimization objective for future stages is to find the possible optimal decision and the corresponding performances. This is the necessary part of the process chain. All these efforts are to find the most optimal performance in the most probable future scenario.

5.5.1 Objectives and Constrains in Present Stage

Minimum total burst number and maximum hydraulic reliability are the objectives in present stage. According to existing research, the entire network's burst number in a specified year (e.g., the next year) can be predicted. The reliability in the following is concerned with the capability dealing with uncertain water demand or some accidents (e.g., fire-fighting). Modified Resilience Index (MI_r) proposed by Jayaram and Srinivasan et al. (2008) is the measurement of network's reliability.

1.   Optimization Objectives

The decision objectives in present stage can be summarized as follows:

Objective 1: Min {total burst number in present decision stage}

Objective 2: Max {hydraulic reliability in present decision stage}

The total burst number is calculated through Eq. (3.11) and hydraulic reliability is measured by Modified Resilience Index (MI_r) through Eq. (5.2). The present decision stage refers to the period after rehabilitation. Because the units of objectives are different, these parameters are to be converted into fitness function through decision maker's judgement and preference. This is addressed in Section 5.6.1.

2. Constraints

The decision constraint in present stage could be summarized as:

Constraint 1: continuity equations and energy equations

Constraint 2: available standard pipe diameter

Constraint 3: budget limit

Constraint 4: minimum pressure (insufficient water pressure)

Constraint 1 is a basic natural law. The hydraulic computation of a network must follow such a law. Constraint 2 is proposed according to pipe manufacture standards. Both Constraints 1 and 2 are rigid constraints and cannot be violated at any time.

Constraints 3 and 4 are the constraints being converted from sub-objectives. They are usually very clear while they have some flexibility in some cases. Because they are set by subjective judgement, the violation in some certain degree is still acceptable in practice.

As for Constraint 3, budget limit is determined by decision maker's affordability and expected WDS's performance. Generally, more budgets bring better performances. The budget could be an absolute or relative value which is a percentage ($\alpha\%$) of total asset value (i.e., total installation cost). Whatever the form it is, the essence is same. The direct cost, which can be classified into loss and expenditure, is subject to the available budget. The loss mainly refers to water loss and the corresponding economic losses. The expenditure generally means the repair, replacement and other renewal action's cost.

Budget in different stages is flexible or rigor. Moreover, the budget in near future is more accurate and certain than that in far future. Usually, the annual available budget is approximately decided in plan. If the budget is flexible, which means the minor cost overrun is allowed, more solutions will be feasible. If the budget is very rigor, the direct cost will be strictly controlled by the available budget. This constraint will make more decision infeasible.

The direct cost should be discounted into net present value (NPV) so that the monetary cost in different years can be compared. Discount rate r is just the bridge to discount the future monetary cost into present cost. The value does not exactly equal to the interest rate. Generally, for public utility works, which takes non-profit as one of the main purposes, the discount rate is relatively low. The lower the discount rate, the total direct cost in the future will be discounted less. Namely, the future economic performance of the system is also important. If an extreme value, $r = 0$, is taken, that means no discount, or the future costs and the present cost is equally important.

In Constraint 4, the minimum pressure requirement is usually determined by consumers' request in WDS development planning. The higher minimum pressure requirement denotes higher service requirement in a node. A suitable and moderate minimum pressure requirement is needed before decision making. Generally, there is a crisp requirement for minimum pressure. It reflects the consumers' benefits around each node and is not ready to be traded-off by higher pressure in other nodes. Although minimum pressure request is relevant to hydraulic performance, it is not the surrogate of that. If the node pressure is above the minimum pressure, the water demand is fully available. Otherwise, the water demand is partly available. Constraint 4 is mainly determined by consumers' tolerance to the insufficient pressure.

If budget constraint or insufficient pressure constraint is violated to some degree, a decision is not necessarily infeasible completely. Some penalty factors can be used to reduce the fitness if some constraints are violated. The general idea of penalty function is imposing a discount by using penalty factor to the fitness function if the flexible constraint is violated severely. If there is no violation, no penalty is needed. Penalty factor is addressed in Section 5.6.1.

5.5.2 Objectives and Constrains in Future Stage

Time span of whole life can be divided into multiple rehabilitation stages. Although it refers to the asset's whole life literally, the time span of whole life is not a determined duration and it is not practical in solving engineering problem. The actual time span often depends on the practical engineering needs and planning horizon. The essential of whole life view is a comprehensive and long-term point of view. In practice, the life span is usually decades of years. With the time extension, the current action's influence become less and less in the future so that it can be ignored gradually. Therefore, a proper time span will be used in practice instead of the theoretical whole life span. In addition, the number of years in each stage is also determined by specified model although rehabilitation is done every year.

An alternative method to deal with the multiple scenarios and performances in multiple stages is needed. As the different uncertainty degrees in multiple stages, the focus of attention for present and future stages should be different. Objectives of decision making are also different. The future uncertain scenarios and the objectives have to be simplified. For decision maker, the present decisions are more practical and more realistic which produce immediate costs and effectiveness. Meanwhile, since there is no absolute optimal decision suitable for various probable future scenarios, it is impossible to find such an ideal and absolute optimal decision for present and future situations. The secondary best goal is to find a set of non-inferior solutions for current situation. Because the non-inferior solutions have equivalent rank in optimal decision making without any other information, each of them will be tested in most probable future scenarios. Namely, the decisions will not only provide non-inferior consequences currently but also a good foundation to deal with most probable scenarios in the future. This is the basic guideline to unify goals of current and future.

1. Optimization Objectives

In order to deal with the multi-objectives and uncertainty, the objectives and decision premises in future stages have to be simplified. The multi-objectives can be combined into one comprehensive objective so that only one optimal solution, instead of multiple near Pareto solutions, will be found in each rehabilitation stage. Due to the unit difference of the two objectives, they are to be converted into fitness and then integrated into one. The single objective is the combined fitness with penalty factors. The decision premise is the assumed most probably scenario, including pipe burst number, water demand and pipe roughness.

This optimal solution/decision will be the premise of further optimal solution in future stages. The uncertain scenarios should be simplified as a certain premise chain, which is the typical scenario followed one by one in estimation. It is assumed that the most probable water demand and breakage number in future are determined values which can be predicted in some models. Under such a certain premise, the WDS's breakage number and node pressure in the future can be accurately calculated. Without these simplifications, the computation premise will become very complex. Moreover, the increasing computation load has little added value for decision making in current stage because the complex decision serials may confuse decision maker.

The optimization objectives in the future stages are essentially the same as those two in present stage but they have to be simplified. The premise for future scenario will be very complicated if all of these non-inferior solutions are preserved. One reason is that multi-objective optimization will generate a group of non-inferior solutions, and each one of which will lead to a new group of non-inferior solutions in next rehabilitation stage. The number of solution serials is similar to the expansion of geometric series. With the stage number increasing, the number of solution serials increases greatly as well. In contrast, single-objective optimization will lead to only one near-optimal solution in each stage. The whole scenario and decision process is clear if one solution is adopted.

2.   Constraints

The constraints in future stages are the same as those four in the present stage. The difference is that the exact value of budget limit and minimum pressure constraint (insufficient water pressure) in each stage might be different. The violations to constraints are converted into penalty factors with same functions.

5.6 Optimization Algorithm for Present Stage Decision

In this study, fast elitism Non-dominated Sorting Genetic Algorithm II (NSGA-II) and a revised NSGA II with mutagenesis will be compared by applying for present stage decision making.

Except for the high evolution rate, a good optimization method should meet the following requirements:

(1) The approximate Pareto solution should distribute uniformly in the near optimal front; and

(2) The approximate Pareto solution should distribute widely in the near optimal front so that each objective can be covered by these approximate Pareto solutions.

The comparison of these algorithms is to find a good method with rapid evolution rate, wide and uniformly distributed near Pareto solutions. The outcome of the decision for current stage will be the foundation of further analysis in future scenarios.

5.6.1 The Application of NSGA II in WDS Rehabilitation

Deb modified NSGA and proposed NSGA II (Deb *et al.* 2000), a fast elitism non-dominated sorting genetic algorithm for multi-objective optimization. NSGA-II implements elitism strategy, effective non-dominated sorting, and protection of parameter diversity, so that the convergence and diversity of the solution are greatly improved (Reed *et al.*, 2013). Among a large number of multi-objective evolutionary algorithms, NSGA II has demonstrated its better performance than some other optimization algorithms in some literature. In some latest literature, genetic algorithm (GA) and its derived algorithms were widely used in optimization research. Genetic algorithm has been used in water distribution leakage analysis. For example, the leakage location and leakage amounts can be estimated through a leak detection method based on EPANET and genetic algorithm (Wang *et al.*, 2012) .

NSGA II overcomes three drawbacks of NSGA:

(1) The computational complexity is reduced from $O(MN^3)$ to $O(MN^2)$ (where M is the number of objectives and N is the population size).
(2) An improved selection operator is provided that creates a stronger mating pool that allows the Pareto front to be approached more quickly.
(3) It is unnecessary to specify a sharing parameter. Instead, the application of crowding distance guarantees solutions' diversity.

The algorithm flow chart of NSGA II is shown in Figure 5.2. An initial population P_0 with a size of N is designed in the beginning. The initial population P_0 is the parent population. After selection, mating, crossover and mutation operation, an offspring population Q_0 is obtained. After merging the population of P_0 and Q_0, the size of new population R_0 is $2N$. For the new population, the individuals are sorted in different levels according to the domination relation. The domination relation and crowding distance determine N individuals to form a new generation P_1. Then, the new generation P_1 is taken as a parent generation. This is an evolution round. Following such an iteration algorithm, the population evolves gradually until the terminating condition is satisfied.

In the sorting of the combined population, the level of non-dominated individuals are assigned as 1. Then all the non-dominated individuals are removed. For all the residual individuals, the new non-dominated individuals are found and assigned a new level number. In the process of new population construction, the individuals of the first k levels from low to high are collected as the new population P_{t+1} if the individual number is less than N. For the individuals in the level of $k+1$, those with low crowding distance are prone to be collected into the new generation until the total number of individuals is N.

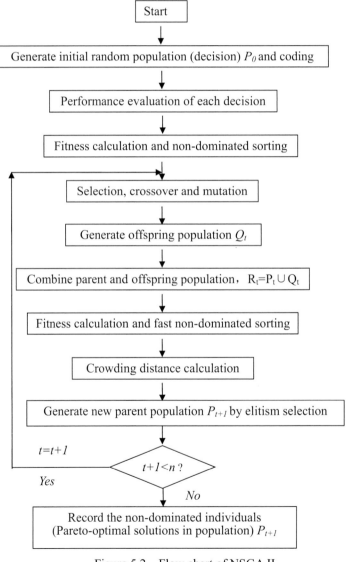

Figure 5.2   Flow chart of NSGA II

In NSGA II, crowding distance is used to estimate the solution density around a solution instead of shared fitness. Figure 5.3 shows the crowding distance calculation. For the density estimation of solutions surrounding a particular point in the population, the largest cuboid enclosing the point i without including any other point in the population is taken. Such a cuboid's two diagonal vertexes are the nearest points of the same rank on either side of this point. The length of diagonal or the average of the two sides is the crowding distance. In Figure 5.3, the crowding distance of the i-th point is the length of cater corner or length of sides of cuboid (shown with a dashed box).

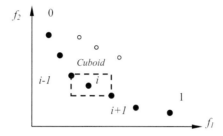

Figure 5.3   The crowding distance calculation (after Deb *et al.* 2000)

In the algorithm, the non-domination rank and crowded distance guides the selection process. This ensures a uniformly spread out Pareto-optimal front (i.e. between two solutions with differing non-domination ranks the point with the lower rank is preferred, otherwise (if both points belong to the same rank), then the point which is located in a region with lesser number of points is preferred). Otherwise, if both points belong to the same rank, then the point which is located in a region with lesser number of points (the size of the cuboid enclosing it is larger) is preferred. Such an automatic adjustment makes the population distribute uniformly.

The following is the explanation of some main content:

1.   Generate Initial Random Population (Decision) and Coding

Pipe criticality index is an indictor to address the urgency or priority rank of a pipe to be rehabilitated. Meanwhile, each pipe's priority is clear because pipe criticality index is crisp. Therefore, it is employed to assist initial decision generating.

In the criticality index calculation, each pipe's rehabilitation approach is determined by the ratio of pipe condition assessment index and significance index after normalization and weighting. If the ratio is greater than 1.0, this pipe is preferred to be replaced because pipe condition dominates the significance. If the ratio is less than 1.0, relining is preferred. If it is just equal to 1.0, either approach is suggested. If flow velocity exceeds the economic velocity's upper limit, which is an empirical value, the pipe is preferred to be replaced by a larger one.

According to pipe condition, significance and velocity, the recommended renewal approach of each pipe can be obtained. Hence, each pipe's renewal expenditure is known as well. Accompanied with pipe criticality index, which pipes are to be renewed and the rehabilitation method for each pipe can be decided until the budget is used up. This is the determined approach to generate an initial individual.

However, the initial population is a group of random decisions instead of one in NSGA II. Therefore, a random approach to generate population is needed. Complete randomly generated solutions might be distributed widely in the set of feasible solutions, but most of them lead to worse performances than that generated from the criticality comparison. If a high select pressure is applied, the better solution will dominate the entire population so that the evolution falls into local optimization. If a low selection pressure is applied, the bad genes are not easy to eliminate and the evolution process will be slow. It is preferred for the population to be randomly distributed around the optimal solutions so that the evolution can converge to the global optimal solutions. In addition, population's diversity is also of concern. If the fitness of the initial population does not vary too much, the competition pressure will not lead to dominate some individuals but occupy the entire solution space. Therefore, the other individuals in the initial population can be generated according to criticality index through some regulations. Roulette wheel selection is one of the widely used methods and is applied in this model. Through Roulette wheel selection, the pipes with high criticality index have more probability to be chosen so that some are selected more than once. After removing the duplicated pipes, the selected pipes can be sorted in descending order of the criticality index. In a similar approach, pipes to be renewed can be determined according to the budget limit. So far, quite a few individuals can be generated through such a random approach. The entire initial generation is composed of the individuals generated through both approaches.

Integer coding is applied as renewal approach code in this model:

0: no action

1: relining

2: replace with same pipe diameter

3: replace with larger pipe diameter

Each renewal decision is an individual in NSGA II. Each gene is the renewal approach code (0, 1, 2 or 3) on each pipe. The total gene digit number is the total water main number. The code in each digit represents the rehabilitation approach for the pipe. For any determined rehabilitation approach, the other features of a pipe, such as diameter, are all determined. For example, an individual (a decision in current stage) in the population might be: 000100030....

2.   Performance Evaluation of Each Decision

For each individual in the population, their performances in various aspects are to be calculated. The performance indicators are modified resilience index (Mlr), total breakage number, insufficient pressure index and total direct costs. For the pipes to be renewed, some of the features are to be changed:

(1) Relining: all of the features are the same as before except for roughness;

(2) Replacement with the same diameter pipe: new pipe age is zero; roughness is changed and can be derived by Eq. (5.4) and Eq.(5.5); historical breakage number is zero.

(3) Replacement with larger diameter pipe: pipe diameter is 50 or 100 mm larger than old pipe; other changes are same as replacement with same diameter.

According to these changes, nodes' pressure can be obtained through hydraulic calculation in each decision. Since the minimum pressure requirement is determined, the modified resilience index of each decision can be derived.

For each group of the same diameter and material, equivalent length, equivalent age and equivalent breakage record can be obtained. Thereafter, each group's breakage number in a specified year in the future will be estimated. Finally, total breakage number and total direct cost of each rehabilitation decision can be derived.

3.   Fitness Calculation

Fitness calculation includes each individual's performance direct fitness and its penalty factors if some constraints are violated. Once penalty factors are involved, a modified fitness can be obtained. According to fitness, non-dominated sorting of this population can be done.

(1) Fitness

Fitness is the conversion of specified objective. There is no authorized or universal function type for fitness. It depends on the specific problem. It is also determined by decision maker's judgment so that the results are consistent with decision maker's bias and preference. Such conversions provide decision maker with proper understanding of the performance.

Fitness 1: the fitness of total break number in a year. A linear decreasing function is applied as the fitness function of total breakage number in a year:

$$f_1 = \begin{cases} 0 & (BR > N_{max}) \\ 1 - \dfrac{BR}{N_{max}} & (BR \leq N_{max}) \end{cases}$$ (5. 10)

Where, BR is the total estimated breakage number in a year, N_{max} is the maximum tolerated total breakage number. N_{max} is a threshold that the decision consequence is completely unacceptable if the breakage number exceed such a limit. In the case of $BR > N_{max}$, the fitness is 0. In an ideal case that total estimated breakage number is 0, the fitness is 1. The threshold value of N_{max} is determined by decision maker. It can also be referred to historical statistics by decision maker. The linear function is easy to understand and suited for judgement.

Fitness 2: the fitness of modified resilience index (MIr), which is the measurement of hydraulic reliability. This index reflects the system's capacity to deal with uncertainty, such as some accidents or unpredicted demand. The decision maker may provide a limit to show the best satisfaction. For example, if the reference ratio is equal to or greater than $\alpha\%$, the fitness of MIr is 1, which means the decision maker is satisfied completely. Therefore, the fitness of MIr is

$$f_2 = \begin{cases} 0 & (MIr \leq 0) \\ \min(1,(\dfrac{MIr}{\alpha})) & (MIr > 0) \end{cases}$$ (5. 11)

In Eq. (5. 11), the fitness is in the interval of [0, 1]. If $MIr \leq 0$, such a decision is completely unacceptable because most of the nodes' pressure is less than minimum requirement and there is no residual energy available to deal with uncertainty. If $0 < MIr \leq \alpha$, there is partly satisfaction or fitness which increase linearly with MIr. If $MIr > \alpha$, the fitness is 1.

(2) Penalty Factor

Although there is no universal penalty function, the function type and the coefficient can be established through the investigation of decision maker.

Here is an example of penalty function creation. Suppose the final fitness is the product of fitness and penalty factor. The penalty factor varies between 0 and 1. Zero denotes the most serious penalty and makes the alternative infeasible. One means the constraint is not violated and no penalty. The general principle of penalty is that the penalty is more severe if the violation is greater. A possible and general function expression could be the following function:

$$p = \frac{1}{1 + a \cdot \Delta^b} = \frac{1}{1 + a \cdot (\frac{x}{X} - 1)^b} \tag{5.12}$$

Where, p is the penalty factor, a and b $(a, b > 0)$ are coefficients that are determined by decision maker, Δ is the excess ratio to the limit, X is the constraint of upper limit, x is the performance indicator (e.g., total direct cost). In the case of direct cost, there is an annual budget limit. Some investigations (e.g. questionnaire) can be done in order to obtain the decision maker's judgement and preference. The questionnaire might be designed as: how much penalty is preferred if the excessive ratio is $(\frac{x}{X} - 1)$? Some typical curve corresponding to some coefficients will be drawn and provided for them to make decision.

Only if the principles are followed, there are other function types to be compared. In the similar investigation, decision maker's preference can be expressed by a proper function and its coefficients.

Penalty factor 1: Budget violation penalty

If budget is violated to some degree, the penalty function is

$$p_1 = \begin{cases} \dfrac{1}{1 + a_1 \cdot \left(\dfrac{Cost^{dir} - Bud}{Bud} \right)^{b_1}} & (Cost^{dir} > Bud) \\ 1 & (Cost^{dir} \leq Bud) \end{cases} \tag{5.13}$$

Where, p_1 is the penalty factor of budget violation, a_1 and b_1 are coefficients, $Cost^{dir}$ is direct cost, Bud is the budget (cost) limit. Generally, the violation penalty factor decreases rapidly from 1.0 with increasing of direct cost if $b_1 < 1.0$. When $Cost^{dir}$ is much more than Bud , the penalty factor may be close to 0 which means the modified fitness is almost close to 0 and the decision is infeasible as well. As the function type is decided, the values of coefficients can be estimated through trial. In the feasible intervals of a_1 and b_1 , some uniformly discrete values can be obtained. Decision maker can choose one of the corresponding curves as the best one to depict his preference.

Penalty factor 2: Insufficient pressure penalty

If some nodes' pressure is below the minimum pressure requirement, the influence scope and degree is subject to the maximum tolerance due to insufficient pressure. In this case, penalty factor is:

$$p_2 = \begin{cases} \dfrac{1}{1 + a_2 \cdot (\dfrac{H_{in}}{H^*_{in}})^{b_2}} & (H_{in} > 0) \\ 1 & (H_{in} \le 0) \end{cases} \qquad (5.14)$$

Where, p_2 is the penalty factor of insufficient pressure, H_{in} is the insufficient pressure index, H^*_{in} is the insufficient pressure index threshold which is determined by water distribution network's planner, a_2 and b_2 are coefficients respectively.

Because the penalty is imposed on the fitness in the form of product, the entire penalty impact is the geometric mean of all the penalty factors. To combine the two penalty factors, the integrated penalty factor $\bar{p}$ is as follows:

$$\bar{p} = (p_1 \cdot p_2)^{1/2} \qquad (5.15)$$

That the geometric mean is used as the average penalty factor is based on the following three considerations:

(1) Each penalty factor is calculated from different dimensions (over budget penalty and insufficient pressure penalty), and is not suitable for arithmetic average.

(2) Each penalty coefficient is probable to approach 0, indicating that the decision is practically infeasible. If arithmetic mean is applied, this extreme and infeasible case may be concealed.

(3) Geometric mean is less affected by extreme value than arithmetic mean.

The revised fitness, which is abbreviated as fitness in the following content, is the product of original fitness and integrated penalty factor.

$$f_1' = f_1 \cdot \bar{p} \tag{5.16}$$

$$f_2' = f_2 \cdot \bar{p} \tag{5.17}$$

Where, $\bar{p}$ is integrated penalty factor, f_1' and f_2' are revised fitness values respectively. Because both the two fitness values f_1' and f_2' are in the interval of [0, 1], the arithmetic mean value of them is a straightforward indicator to reflect the mean fitness of a decision.

$$fit = \frac{1}{2}(f_1' + f_2') \tag{5.18}$$

Where, fit is the mean fitness of a decision.

4. Selection

The random selection approaches in GA usually include roulette wheel selection, random walk sampling, local selection, tournament selection and so on. After the individual's fitness is assigned, a suited selection approach can be chosen. In this study, roulette wheel selection is applied. The fitness will be converted into an interval on the roulette, the formulation is as following:

$$P_k = \frac{f_k}{\sum_{j=1}^{n_p} f_j} \tag{5.19}$$

Where, f_j is the mean fitness of the j-th individual, n_p is the population size, P_k is the probability of the k-th individual being selected. The fitness of each individual (chromosome) is converted into an interval on a roulette wheel which is the probability of being chosen through Eq.(5. 19). The individuals with better fitness have more probability or chance to be selected so that the better genes are preserved to the next generation. For each two consecutive roulette wheel samplings, they constitute a pair of parent chromosomes. Only if the two parent chromosomes are not the same, crossover is carried out.

A chromosome's fitness can be expressed by mean fitness fit , breakage number fitness f_1' and modified resilience fitness f_2'. Each of them addresses a chromosome's fitness from different aspects. The parents being selected have some distinctive characteristics according to the three fitness indicators. The different fitness functions are carried out alternately in parents' chromosomes selection so that the population has better diversity. Some good genes are not easy to be removed in such a method. Different selection pressures can be applied in the population selection before crossover. Generally, the whole selection process should consider the balance of evolution rate and population diversity.

(1) In the earlier evolution stages, the evolution rate is more emphasised so that a high selection pressure is preferred to remove some bad individuals.

(2) In the middle evolution stages, the selection pressure should not be as high as that in the earlier stage so that some individuals with moderate fitness have more opportunity to be selected.

(3) In the late stages, the fitness difference is not as significant as the stages before. Less selection pressure is preferred to improve diversity.

In practice, the selection pressure of power function type can be changed in different evolution stages. If the power value is big, the selection pressure is also big and the best individuals have more opportunity to be chosen.

5. Crossover

Crossover is a key step of evolution through which more new individuals can be generated. The searching scope might be enlarged. Crossover probability tells how frequently crossover will be performed. In some literature, the general suggested interval is [0.5~0.95] or some interval around that. If the crossover probability is low, it means the propagation probability is low and a new individual is difficult to emerge.

The crossover probability is 100% in this research. Because all the offspring and parent generation are put into a pool to select better individuals to form new generation in NSGA II, such an operation means all the offspring are made through crossover. Uniform crossover is applied in this study and crossover points can be randomly chosen.

6. Mutation

Mutation probability addresses how frequently the genes in chromosome will be mutated. Mutation is applied (infrequently) to avoid the GA converging on local optima. For any particular off-spring, if mutation is not applied than the offspring stays the same, whereas if mutation is applied, then part of chromosome of the offspring is changed (in a random way).

In this study, all the genes have uniform probability of mutation. For each gene, a random number between 0 and 1 is generated. If this number is less than a threshold, which is the mutation rate, the gene on this location is to be changed randomly among 0, 1, 2 or 3. These codes represent the rehabilitation action approach, including no action.

7. Elitism Selection

In order to expend searching scope, N parents and N offspring are combined to constitute a new population with the size of $2N$. For each of their chromosome, which is a set of genes (decisions) on each pipe, new asset information and corresponding performance can be derived. Before further calculation, the duplicated individuals with the size of n are to be removed. Hence, there are ($2N-n$) individuals left. The removal of duplicated individuals in time can save computation time.

The next step is to sort these non-duplicated individuals, which is a significant step in NSGA II. The two modified fitness on total breakage number and MIr respectively are the indicators of sorting. The primary basis of sorting is the domination level and the secondary basis is crowding distance.

Non-inferior solutions can be found according to individuals' modified fitness. These individuals are the member of elitisms. Then these elitisms from the original collection are moved into an elitism set. If the number of these non-inferior solutions is less than required population size N, more individuals are to be selected from the residual population. Repeat the searching and outputting process until the number of elitism is more than N. For the non-inferior points in the last round, crowding distance is applied to find the individuals.

According to the distance sorting, the individuals with greater crowding distance are prone to be selected until the total elitist number is N. This dominant population is the new generation after selection. Hence, population is updated.

5.6.2 Modification of NSGA II

Random mutation in GA is a key factor to generate new individuals. The direction of purely random mutation is unknown and uncontrollable. Mutation operations have been demonstrated as efficient for solving multi-objective optimal problems (Moosavian and Lence, 2017). In this research, induced mutation (mutagenesis) is proposed. It provides a guidance to direct the evolution by targeting to change some bad genes with higher probability (i.e. mutation probability rate) so as to accelerate evolution rate and generate better individuals.

The key issue of mutation with induction is to find the immediate relationships between rehabilitation action and optimization objectives. In the revised NSGA II, the pipe with excessive flow velocity is prone to be replaced by a larger pipe in some probability, or the pipes with costly rehabilitation expenditure have a little more chance to be renewed in a cheaper approach. This approach is expected to reach the Pareto optimal solution in far less number of generations. Only in this way, is the correct direction of induced mutation to be ensured. Excessive flow velocity always makes high water head loss and lead to some insufficient node pressure. Pipe flow velocity and cost will be the clear inducing factors for mutation. If one rehabilitation decision is penalized due to insufficient pressure, some genes (i.e. actions on some pipes) leading to excessive velocity in this chromosome might be changed by induced mutation to avoid or mitigate the penalty. For example, the diameter might be enlarged. The same for the direct cost. Both of them bridge rehabilitation action and performance immediately.

The novel induced mutation changes the mutation probability and direction to change some bad genes. Namely, the bad genes leading to some bad performance have more probability (i.e. mutation rate) to be changed. In addition, once they are selected, the mutation will lead to performance improvement. In the revised NSGA II, the rules of induced mutation are as follows:

(1) Velocity induced mutation: the pipes with excessive flow velocity have more probability to be replaced by a larger pipe.

(2) Cost induced mutation: if the total direct cost violates the budget constraint, the pipe with

maximum renewal expenditure has more probability to be renewed in a cheaper rehabilitation approach.

(3) Velocity and cost induced mutation: both of the two factors work alternatively in the whole evolution process.

Velocity induced mutation is applied to increase the mutation probability that the pipes with excessive flow velocity are to be chosen. Once a pipe is chosen to be replaced because of its high flow velocity, the new pipe's diameter should also increase. For similar reasons, the pipes to be replaced or relined with expensive cost are prone to be rehabilitated by a cheaper method or even no rehabilitation action on it at all. Although this affects the performance, the cost is lower and the modified fitness might be improved. Note that the induced mutation is an auxiliary approach of random mutation instead of an independent mutation approach. In order to keep the total mutation rate same as that before modification, the mutation rate of the randomness and that for the excessive velocity (or high cost) is half of the total mutation rate respectively. In addition, alternatively using all three approaches (one random mutation and two induced mutations) will also be attempted.

Mutation with induction will work with random mutation alternatively to avoid over-emphasising the influence from the induction factors. In this study, induced mutation is expected to accelerate the evolution rate but not influence population diversity, especially in the early evolution stages.

In the trial computation, it is found that most of the computation time is for hydraulic calculation, instead of evolution operations, in a round of evolution. However, hydraulic computation is necessary in each round of evolution. Hence, it is important to reduce iterations. By induced mutation, it is expected to reduce computation load and accelerate evolutionary rate. This modification is expected to reach the near Pareto front with far less number of generations.

There is no absolute standard for stopping iteration. One possible indicator is the generation number which is derived from some trial computation. The generational process is repeated until a termination condition has been reached. If one terminating conditions is satisfied, the evolution computation might stop. The terminating conditions usually are when:

(1) Fixed number of generation is reached;

(2) The highest ranking solution's fitness has reached a plateau and successive iterations will not improve it;

(3) Manual inspection; and

(4) Combinations of the above.

Because there is neither minimum criterion nor prior experience about the termination condition, a fixed number of generations are derived after some trial computation.

5.7 Optimization Algorithm for Future Stages Decision

For the present decision stage, a group of near-optimal solutions/decisions for multi-objective are to be obtained. Based on each of these solutions, the future performances and corresponding rehabilitation actions in future stages will be simulated so that the long-term influence of each current decision can be tested. The current renewal decision is the premise and foundation of further rehabilitation. This is the link of the decisions between present and future stages.

Due to the uncertainty in future, there is no absolute optimal decision for all possible future scenarios. Because any optimization must be corresponding to a specified premise, one decision cannot be always optimal for the uncertain future scenarios. An alternative is proposing a decision which could meet the most probable scenarios as representative in future stages. This is the compromise of the optimistic and pessimistic prediction. Generally, the most probable scenarios are typical scenarios, and also the compromise of all possible scenarios. The decision based on such a moderate scenario is thought to be adaptable to any possible scenarios. If the most probable scenario occurs, such a method will bring the just optimal decisions. Otherwise, the expectation value of wasted or supplementary cost is also not too much. In such a simplification, the future scenarios are predictable and not uncertain any more.

5.7.1 WDS Rehabilitation Decision Process for Future Stages

The optimization objectives of future stages are essentially the same as those of current stage but they are simplified into single objective optimization problems to deal with the great uncertainty in future. Meanwhile, the constraints are converted into penalty factors to play a role. When whole life is divided into some stages, the deterioration and rehabilitation repeat one after another as a chain serial. Because the decision in future stage is based on single objective optimization, the optimal decision is the only one based on the decision consequence in previous stage. The calculation process is shown in Figure 5.4.

The key steps are as follows:

1. Input initial conditions and present optimal decisions

The basic initial condition and some necessary requirement, such as budget limit and minimum pressure requirement, are inputs. In addition, the non-inferior decisions derived from current situation are input and ready as the initial condition. The number of initial non-inferior decisions is k. Each of these decisions will be treated as initial condition for the next analysis.

2. The decision's consequence in the $j-th$ stage

The optimal renewal decision in the beginning of first decision period is made according to present stage situation. According to current renewal decision, the consequence in the following rehabilitation period can be derived. This includes new asset information after rehabilitation and estimated total breakage number in this period after deterioration. A renewal decision's consequence in the end of the first renewal period forms the initial condition for the following decision.

3. Single objective (GA) optimization decision

According to the performance in the previous period, a new single objective optimization decision can be made. Its performances, such as breakage fitness, MIr fitness, and combined fitness are delivered.

4. The consequence in the (j+1)-th stage

According to the optimal decision, the deterioration consequences during or at the end of the stage   are also estimated, such as total breakage number, total direct cost and new asset features (e.g. pipe age, roughness, pipe condition).

Repeat the 3rd and 4th step until all of the rehabilitation period is experienced. Each stage will deliver a near optimal rehabilitation decision and its corresponding performance which will be used as the premise of deterioration and next stage's decision. After all of the rehabilitation stages are experienced, a decision chain is obtained coupled with the modified fitness in each stage.

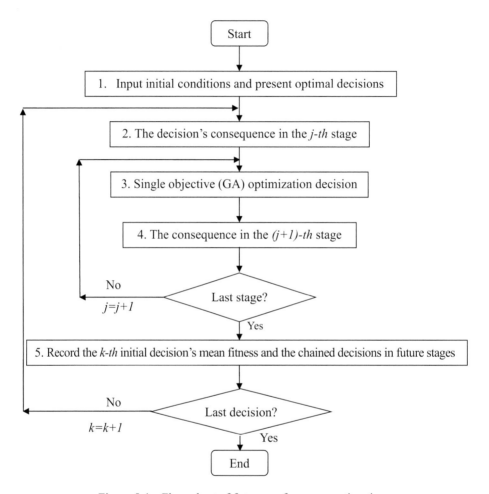

Figure 5.4   Flow chart of future performance estimation

5.   Record the *k-th* initial decision's mean fitness and the chain decisions in future

After the chain optimal decision is made based on the *k-th* initial optimal decision, all of the decision's fitness and the chain decisions in future will be recorded. The breakage number and *MIr* fitness after modification are still the two dimensions to judge a decision's performance in the future. Once all of the present initial optimal decisions（with the number of *k*）are tested in the future, the fitness of each present optimal decisions can be compared.

5.7.2 The Application of Single Objective Genetic Algorithm (GA)

In each future stage, one single objective near optimal decision should be found under some certain premise and the decision in previous rehabilitation stage. The inputs are some initial conditions and requirements. The outputs are one near optimal decision and its corresponding performance and fitness. GA is applied in this optimization with the objective of maximum mean fitness. Figure 5.5 describes the steps of "Single objective optimization decision.

The main steps are as follows:

1. Input initial conditions and generate initial population

The initial conditions include some pipe features after previous pipe rehabilitation and deterioration, such as diameter, age, roughness and so on. The other service requirement (e.g. minimum pressure, budget limit) are also known according to the system's development plan. The initial population generation approach is the same as that in the present stage.

2. Performance and fitness calculation

According to the rehabilitation decision, the performances (i.e. break number, Mlr, direct cost and insufficient pressure index) and fitness can be calculated. The modified fitness is the combination of modified breakage fitness and Mlr fitness which involve penalty factors of insufficient pressure and excessive direct cost. The difference is the modified fitness which is a one dimension parameter. Hence, the best decision in the population which has the highest modified fitness can be found and recorded.

3. Selection, crossover and mutation

Roulette wheel selection is applied to choose parents individuals. The basic rule is the same as that in Section 5.6.2 but the selection pressure might be different. The crossover probability varies in a wide interval with the upper limit of 1.0. Because the best solution in each round of iteration is always preserved, the fitness of each output generation must be equal to or better than that in the previous generation even if all the chromosomes are worse after mating.

The regulation of mutation is also the same as that in Section 5.6.2. The mutation can be one of the following: completely random mutation, velocity induced mutation, cost induced mutation and the combination of velocity and cost mutations. A gradually reduced mutation rate is suggested to be applied. The best chromosome and the offspring generation after mutation are combined for further analysis.

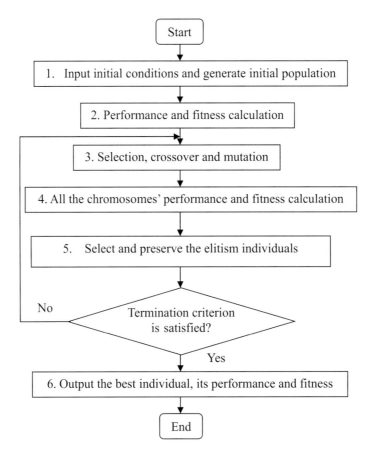

Figure 5.5 Flow chart of single-objective Genetic Algorithm

4. All the chromosomes' performance and fitness calculation

The calculation approach is same as the previous but the process will repeat for all of the best chromosome and the offspring generations in last step.

5. Select and preserve the elitism individuals

According to the fitness of each chromosome, the duplicated ones can be deleted. After sorting the remaining individuals, some elitism individuals which have high mean fitness will be found and preserved as parents for next generation.

Once the termination criterion is satisfied, the evolution will stop. The best individual, its corresponding performance and fitness will be delivered. Otherwise, go back to Step 3 and continue the evolution. The termination criterion might be a pre-set maximum generation number or the fitness difference of the last generations.

6. Output the best individual, its performance and fitness

As the result of single objective optimization calculation, the optimal decision (individual) of this stage is obtained. Its performance and fitness is preserved and output as well.

5.7.3 Final Decision Making

According to present initial condition and some service requirement in development planning, a group of two-objectives near optimal rehabilitation decision can be derived. With the constraints being considered, these decisions are equally non-inferior to each other. Based on these present rehabilitation decisions, the water distribution system's performance in long-term future stages can be simulated. Because the fitness is an integrated one and penalty factors have been imposed on it, the integrated fitness is comparable. The average values of fitness during these multiple future stages are representatives.

Although only one single-objective optimal solution is found in each future stage, it is still can be measured by two fitness dimensions (i.e. breakage number and modified resilience index) with penalty factor. Then, the average value in each of the fitness dimension can be obtained. For example, there are ten future stages in a model. Thereafter, the average fitness of break number with penalty can be calculated for a rehabilitation decision in these ten stages. Such an average value represents the breakage number fitness in all of the future stages. The same job can be done for modified resilience index fitness. Hence, a decision's future fitness in these two dimensions can be generated. Therefore, the decisions which are both the near Pareto points in present and future stages are chosen as the optimal solutions.

5.8 Discussion

The theoretical development of optimization technology and its application in water distribution system analysis have not stopped. The new optimization algorithms either make the efficiency of optimization higher, or make calculation more stable, or have more ability to avoid premature(i.e., stay near the local extremum). Nevertheless, optimization technology has not been widely applied in the production of water supply industry. Walski (2014) pointed out that one of the reasons is that it is easy to obtain results but difficult to determine whether the results are optimal, and another reason is that the high complexity of large-scale water distribution networks make optimization technology impractical (Walski 2014).

The understanding of optimization results is more flexible and practical. In fact, any optimization model is carried out on a specific premise, and the influence of some minor factors is neglected. Some of the actual network conditions are not consistent with the model's premise. Some neglected factors may have a huge impact on the model's results. Therefore, it is necessary to make a more comprehensive analysis of the results calculated by the model, so as to avoid omitting some valuable results.

The output of the model is not an absolute optimal decision because there is no absolute optimal decision in the uncertain premises. The uncertainty widely exists in present decision premise and future development prediction. Enumeration method is not feasible because of the great uncertainty in the future multiple stages and the huge computation load. Therefore, there is possibility of missing some potential optimal solutions if the actual development is different from that in the model prediction.

One question might be: are the decision serials useful if the actual scenario is different from those in prediction? This case is very likely to occur. What is needed is a "no-regret" or "low-regret" decision, instead of a perfect decision. Namely, the decision is relatively optimal under present and future deterministic conditions. Even if the present decision is not very suitable in future development test practice, there is still change to restore the faults in future decision making. The decision is made stage by stage, instead of once and for all. Decision making in each stage is through modelling which is based on new development situation and future prediction. Although there is probability of missing some optimal solution if some atypical scenarios happen, decision maker still have chance to restore through making another optimal decision in the future. This is a "low-regret" approach.

Since it is inevitable to miss some potential optimal decisions in practice in a whole life costing view, a further question might be: is it possible to reduce the regret? The answer is positive but more computation load is needed. In this model, some feasible but not optimal decisions in present stage could be adopted as premises for further options. For example, 20 feasible solutions are found, of which 15 are non-inferior decisions. In the original thought, only the 15 non-inferior decisions will be further tested in the future scenarios and the balance 5 will be neglected. However, all of the 20 decisions can be tested in the future scenarios in order to reduce the possibility of losing some potential good decision in a long term view. It is suggested that all the solutions, inferior and non-inferior, are to be preserved for future scenario tests. Some relative inferior decisions in the present stage maybe bring good performances in the future. Because the population (solutions) in NSGA II and GA generally have a good fitness after evolution and the fitness difference is not great, all of the performance of the solutions might be tested in future scenario.

Although such an approach involves more feasible solutions, they are still far less than all the feasible solutions So, it is only believed that with more feasible solutions involved, it is more probable to preserve the potential optimal solutions. Meanwhile, the computation load increases as well. Therefore, unlimited increasing feasible solution amount is not a good approach. In optimization calculation, the feasible solutions in present stage, which are also near optimal solutions, can be categorized into non-inferior and inferior solutions. In practice, some inferior solution is only slightly inferior to the nearby non-inferior solutions. It is possible that these inferior solutions' performance is better than that of some non-inferior solutions in the future scenarios. For example, in the case of GAs (Genetic Algorithms), the individuals (i.e., decisions) in final evolution generation are these feasible solutions and all of them can be tested in future scenarios. The computation load does not increase too much and more feasible solutions are involved. After testing, the final solutions can be categorised into four groups:

(1) Non-inferior solutions both in present and future scenarios.
(2) Inferior solutions both in present and future scenarios.
(3) Non-inferior solutions in present stage but inferior in future scenarios.
(4) Inferior solutions in present stage but non-inferior in future scenarios.

The decisions in above Group 1 are undoubtedly most preferred because their performance is always non-dominated. Those in Group 2 are not preferred because their performance is always dominated. As for decisions in Groups 3 and 4, decision makers should judge which decision can be chosen based on the trade-off between present and long-term. The solutions in

Groups 3 and 4 are not to be ignored because they still have better fitness either in current stage or future stages. Other factors beyond the model have to be introduced to help decision maker. The evaluation criteria reflect the decision maker's preference.

If the decision maker is conservative, whose concern is more on current existing costs and benefits and certain situation, Group 3 will be preferred. The current performance of the decisions in Group 3 is better than those in Group 4. If the decision maker prefers risk, who likes to obtain more benefits in some probable future scenarios, Group 4 will be preferred. The future performance of the decisions in Group 4 is better than those in Group 3.

In addition, although universality and uniformity are the requirement of the solution distribution, it is not easy to realize it in this model. The main reason is that the decision variables are discrete rather than continuous.

5.9 Summary

This chapter presents an innovative whole-life cost water main optimal rehabilitation decision model.

The main research findings are as follows:

1. Combining Present and Future Stages within a Whole Life framework

Under present situations, several near non-inferior optimal decisions can be found through optimization. The potential influence of each of the decisions is predicted under the most probable scenarios of multiple future stage development, by undertaking optimisation across the future stages. The optimization approach applied for present stage is different from that for future stages, as for the future uncertainty needs to be considered. Due to the uncertainty and chain structure of future decisions, the optimization objectives and constraints are different from and more complicated than those used for present stage. Therefore, the model is simplified by considering only some typical (probable) scenarios and the two optimization objectives are combined into one. The final decision making is based on the current performance after rehabilitation and their future possible performance. Only those decisions that provide improved performance both in current and future stage will be preferred. The steps taken for integration are feasible with respect to computational load.

2. Development of Objectives and Constraints

Among the diverse optimization objectives, the primary objectives are to minimize general total costs (i.e. direct, indirect and social costs) and to maximize benefits.

In the present decision stage, these primary objectives are converted into two objectives and two sets of constraints. One objective is to minimize pipe burst number, which represents minimize indirect costs and social costs, and maximize serviceability. Another objective is to maximize hydraulic reliability, which denotes the network's capability of dealing with uncertain water demand and accidents. The constraints developed are the hydraulic continuity and energy equations, available pipe sizes, annual budget limit and minimum pressure requirement. Direct cost is considered as a constraint (rather than an objective) and so the model seeks to maximise benefits while utilising the full allocated resources for rehabilitation. For the future rehabilitation stages, the objectives are combined into a single objective, where the decision premise is the assumed to be the most probable scenario.

3. Modified Optimization Algorithms for Different Stages

In this chapter a modified NSGA II approach is applied to the water distribution system's optimal rehabilitation decision model. NSGA II is a feasible approach searching for the present stage optimisation where two-objective are considered. A modified NSGA II is developed that has an induced mutation that allows an accelerated evolution rate mainly in the early stages of evolution. Induced mutation in NSGA II with velocity, or cost, or both as the inducing factor is shown to accelerate the evolution. For the future stage optimisation, a single-objective GA is applied to generate optimal rehabilitation actions in the future stages. Only the solutions which are non-inferior both in current stage and future stages are chosen as rehabilitation decisions. This is to guarantee that the decisions are not only the near optimal solution for the present stage but also their applicability and potential performance in the simulated future stage.

4. Understanding of Optimization Results

The output of the optimization model is not an absolute optimal decision as there can be no absolute optimal decision under the uncertain premise. The understanding of the generated optimization results needs to be more flexible, as the actual development may differ from that the model prediction. Hence, the near optimal decision can be viewed as a "no-regret" or "low-regret" decision.

Chapter 6   Case Study

6.1 Introduction

This chapter aims to illustrate the utility of the models developed and presented in Chapters 3, 4 and 5. It applies the pipe breakage number prediction model, the pipe criticality assessment model, and optimal rehabilitation decision model to case studies from the UK. A sub-zone of a large network is used to illustrate the applications of all the models.

The pipe breakage number prediction model utilises numerous pipe data, including information related to the pipes assets and pipe failure records. For the presented case studies, the data (i.e. pipe diameter, material, installation year, length and the total number of bursts recorded during the 9-year (from 2001 to 2009) was obtained from the water utility's database. Other environmental data was obtained from public domain databases including UK weather websites (http://www.weatheronline.co.uk/). The pipe criticality assessment model is applied to a section of a large WDS that is used in the application of the pipe breakage number prediction model. The choice to use a selection of the model is because the pipes' data is missing for some parts of the network. The results of the application of the pipe condition assessment and pipe significance index models are used by the pipe criticality assessment model. The same zone of the WDS is used to demonstrate the optimal rehabilitation model.

The entire application demonstrates how the combination of the models can assist decision makers, faced with a multi-objective optimisation problem, to meet their present optimization requirements, while positively influencing future decisions.

6.2 Case Study of Pipe Breakage Number Prediction Model

In order to simplify the calculation and highlight the main factors, only one pipe material is taken as the object in this case study. Cast-iron pipes are selected because this material has been widely used in practice for years. Both the asset and failure data of cast-iron pipe are relatively abundant. For the other materials, the deterioration principle is quite similar as that of cast-iron pipe.

The entire network's basic data statistics are shown in Table 6.1. This table shows that the number of failures recorded during the 9-year monitoring period is approximately 8% of the total number of pipes. The failure record in each year is very rare. Because the failure records from 2001 to 2002 are too few, it is suspected that the record is incomplete and the data in these two years are not adopted in the case study. Therefore, the valid failure data is from 2003 to 2009.

Table 6.1Database source overview (only for cast-iron pipes)

Features	Values
Year of installation	1890-2003
Year of failure record	2001-2009
Nominal diameter	75 mm-450mm
Total length	792 Km
Number of cast-iron pipes	16057
Number of failure records	1316

From the entire asset database, about 20% of pipes are randomly selected as the testing data and the balance 80% are taken as training data. Table 6.2 is the overview of these data groups. Because the pipes with a diameter less than 50mm or the pipes failed in 2001 are quite few, the actual representativeness (i.e. year of installation, year of failure record and nominal diameter) of the two groups (80% and 20%) is quite the same. The total pipe length, number of pipes and failure record follow the appropriate ratio of 20% and 80% to the total quantity respectively.

Table 6.2 Database source overview for training and testing (only for cast-iron pipes)

Features	Values (80% data for training)	Values (20% data for testing)
Year of installation	1890-2003	1890-2003
Year of failure record	2001-2009	2001-2009
Nominal diameter	75mm-450mm	75 mm-450mm
Total length	632Km	160 Km
Number of cast-iron pipes	12885	3172
Number of failure records	1044	272

If a water main's record is not completed (e.g. there is no installation/replacement year on record, or year of repair is unknown for a repaired pipe), such a pipe's record should be removed. Such a pipe's record is not helpful to the analysis because both pipe age and failure time are key factors in this study. The amount of such deleted data is few. In addition, some small pipes (i.e. diameter less than 50mm) data are removed as well because the failure mechanism and the environment of them is different from that of mains.

6.2.1 Data Classification and Aggregation

1. Basic Classification

As described previously, asset data are classified into basic groups by the material, diameter and age. In this research, only cast-iron pipes are selected. All the pipes have been grouped into 11 diameter classes (i.e., 75mm, 100mm, 125mm，150mm，175mm，200mm，225mm，250mm，300mm，375mm and 450mm).

The latest installation record is in 2002. The raw asset and incident data are stored in different layers of a geographical information systems (GISs) database. The valid data are extracted in GIS format and transformed into Microsoft Excel format.

Table 6.3 is an example of one of the basic groups for model training. It displays the classified break record for all pipes of 75 mm and at the age of 1-year during the entire monitoring horizon. Because the effective monitoring horizon is from 2002 to 2009, the year of installation/replacement for the 1-year-old pipe vary from 2001 to 2008 accordingly. Such a table clearly records total installation/replacement length and breakage number for a specified pipe group that have the same pipe age and same diameter simultaneously. All the asset data are classified in such a basic group for which the equivalent age, equivalent diameter, failure number and renewal length in each year are recorded.

Table 6.3   Breakage number and asset statistic result of 1-year old pipe (75 mm)

Year of installation	2001	2002	2003	2004	2005	2006	2007	2008
Year of observation	2002	2003	2004	2005	2006	2007	2008	2009
Installation/replacement length (m)	5973	277	0	0	0	0	0	0
Breakage number	2	0	0	0	0	0	0	0

For such a basic group, the recorded average breakage rate (breakage number per unit length per year) can be calculated as follows:

$$Br_{D,Age} = \frac{\sum\limits_{t=1}^{T} \sum\limits_{i=1}^{n} BR_{D,Age}^{i}}{\sum\limits_{t=1}^{T} \sum\limits_{i=1}^{n} L_{D,Age}^{i}}$$

(6. 1)

Where, $Br_{D,Age}$ is the average breakage rate (breakage number per unit length per year) for the pipe group with diameter of D and age of Age, n is the total pipe number in this group, T is the total observation (record) period, $BR_{D,Age}^{i}$ is the i-th pipe's breakage number in the group at the age of t, $L_{D,Age}^{i}$ is the i-th pipe's service length in such a group at the same age. In the example of Table 6.3, $\sum\limits_{t=1}^{T} \sum\limits_{i=1}^{n} BR_{D,Age}^{i}$ is the sum of the row of "Breakage number" , and $\sum\limits_{t=1}^{T} \sum\limits_{i=1}^{n} L_{D,Age}^{i}$ is the sum of the row of "Installation/replacement length" .

2.   Further Aggregation

In this case study, there are 91 aggregated groups, of which 80 are aggregated by the same age and 11 by same diameter. Pipes older than 60 years are rare and their breakage records from 2001 to 2009 are very few as well. In order to avoid the wrong analysis and judgment of the general pipe breakage tendency, the data of these very old pipes are removed in this case study. Table 6.4 shows the data grouped by the same age and Table 6.5 shows the data grouped by equivalent diameter. The failure number and equivalent length (i.e. total length) vary greatly in these groups. Especially, most of the groups' failure number is zero when the age is above 40 years in this case study.

Freezing index is derived from historical weather records. The average freezing index from 2003 to 2008 is 7.48℃-day in this case study.

Table 6.4   Data grouped by age

Age_{class} (Year)	D_{class} (mm)	L_{class} (m)	Br_{class}^{re} (10^{-2} break/year/km)	Breakage number (2009)
1	112.2	2457	5.09	0
2	111.8	2509	0	0
3	115	2600	4.81	0
4	119.8	4116	12.15	0
5	124.3	5050	17.33	0
6	114.8	11251	4.44	0
7	108.8	17654	2.83	0
8	109.2	28178	3.55	1
9	105.8	39028	4.16	3
10	105.3	48313	8.54	0
11	103.9	53345	11.95	0
12	103.5	57246	14.63	0
13	103.8	56855	23.74	1
14	103.6	56170	22.92	0
15	103.6	51565	25.45	2
16	103.6	45804	24.56	4
17	104	35328	23	12
18	108.7	25550	21.04	23
19	103.5	16003	21.09	16
20	108.9	10977	15.94	13
21	116.6	7053	19.49	8
22	119.8	6209	30.2	6
23	116.5	6042	26.9	1
24	109.1	4338	25.93	0
25	116.3	3397	18.4	0
26	121.6	2911	21.47	1
27	132	1788	6.99	0
28	118.5	1721	14.53	0
29	184.2	6317	3.96	1
30	172.2	7413	1.69	0
31	180.3	7615	8.21	0
32	178.9	7545	4.97	0
33	180.9	7751	14.51	0
34	182.8	7784	12.85	0

Age_{class} (Year)	D_{class} (mm)	L_{class} (m)	Br_{class}^{re} (10^{-2} break/year/km)	Breakage number (2009)
35	189.8	7118	12.29	0
36	189.6	6837	18.28	2
37	212.5	5605	4.46	0
38	278.1	2879	13.03	10
39	298.9	793	0	0
40	293.8	691	0	2
41	239.9	842	0	0
42	169.2	469	0	0
43	178.1	422	0	0
44	181.7	406	0	0
45	108.1	774	16.15	0
46	108.6	762	0	0
47	108.6	760	16.44	0
48	108.1	760	0	0
49	108.1	761	0	0
50	122.8	531	0	0
51	116.6	611	0	0
52	116.6	611	0	0
53	81.8	567	0	0
54	75	114	0	0
55	75	114	0	0
56	75	114	0	0
57	75	94	0	0
58	84.2	107	0	0
59	84.2	107	0	0
60	111.1	27	0	0

Table 6. 5   Data grouped by diameter

D_{class} (mm)	Age_{class} (Year)	L_{class} (m)	Br_{class}^{re} (10^{-2}break/year/km)	Breakage number (2009)
75	14.9	285180	15.95	55
100	15.52	248922	20.59	41
125	12.39	6491	19.26	1
150	16.56	83900	9.09	6
175	12.78	2196	34.15	0
200	12.31	2611	19.15	1
225	12.72	10305	2.43	1
250	17.98	6604	0	1
300	25.74	14990	0.83	0
375	19.04	5042	0	0
450	30.56	17544	3.56	0

6.2.2 Formula Type Selection and Weighted Nonlinear Regression

Due to the pipe length variety of different pipe groups, the contribution of each group error to the total error is not the same. Hence, the weights derived from pipe length are needed in regression. The weights are derived from Eq. (3.15). These weights are used to modify the data after aggregation.

According to existing research (e.g., Berardi *et al.* 2008), power and exponential functions are usually carried out for each influence factor. If each influence factor has two simple function forms (power or exponential function), and a factor's positive or negative influence is determined, there are $2^4 = 16$ alternative function types for these four factors. It should be noted that $f_2(FI_{class}) = 1$ in Eq. (3.11) and Eq. (3.12) if $FI_{class} = 0$. This is because the temperature has no effect on the pipe deterioration if there is no freezing. The coefficients and errors comparison can be calculated through the software called "1stopt". Table 6.6 shows the mean square error (MSE) and residual sum of squares (RSS) of all the formula types in trial. The results show that all the formula types have quite good and similar correlation coefficients. Therefore, the mean square error and residual sum of squares determine the selection. Through the comparison of the 16 function types, the 11th type in the table is suggested due to its low mean square error and residual sum of squares. The correlation coefficient between the recorded and estimated values is 0.903. Table 6.7 lists the coefficient values in the *11th* equation type (using 80% of the data).

Table 6.6   Possible formula types (16 types)

No.	Function	Mean square error (MSE)	Residual sum of squares (RSS)
1	$BR_{class}^{as,T} = a_0 \cdot I_{class} \cdot (Age_{class})^{a_1} \cdot (FI_{class}+1)^{a_2} \cdot (Br_{class}^{re})^{a_3} /(D_{class})^{a_4}$	0.608	26.27
2	$BR_{class}^{es,T} = a_0 \cdot L_{class} \cdot \exp(a_1 \cdot Age_{class}) \cdot (FI_{class}+1)^{a_2} \cdot (Br_{class}^{re})^{a_3} /(D_{class})^{a_4}$	0.684	33.24
3	$BR_{class}^{as,T} = a_0 \cdot L_{class} \cdot \exp(a_1 \cdot Age_{class}) \cdot \exp(a_2 \cdot FI_{class}) \cdot (Br_{class}^{re})^{a_3} /(D_{class})^{a_4}$	0.689	33.69
4	$BR_{class}^{es,T} = a_0 \cdot I_{class} \cdot \exp(a_1 \cdot Age_{class}) \cdot \exp(a_2 \cdot FI_{class}) \cdot \exp(a_3 \cdot Br_{class}^{re}) /(D_{class})^{a_4}$	0.707	35.45
5	$BR_{class}^{es,T} = a_0 \cdot I_{class} \cdot \exp(a_1 \cdot Age_{class}) \cdot \exp(a_2 \cdot FI_{class}) \cdot \exp(a_3 \cdot Br_{class}^{re}) / \exp(a_4 \cdot D_{class})^{a_4}$	0.699	34.66
6	$BR_{class}^{es,T} = a_0 \cdot I_{class} \cdot (Age_{class})^{a_1} \cdot \exp(a_2 \cdot FI_{class}) \cdot (Br_{class}^{re})^{a_3} /(D_{class})^{a_4}$	0.699	34.66
7	$BR_{class}^{es,T} = a_0 \cdot L_{class} \cdot (Age_{class})^{a_1} \cdot \exp(a_2 \cdot FI_{class}) \cdot \exp(a_3 \cdot Br_{class}^{re}) /(D_{class})^{a_4}$	0.628	28.00
8	$BR_{class}^{es,T} = a_0 \cdot L_{class} \cdot (Age_{class})^{a_1} \cdot \exp(a_2 \cdot FI_{class}) \cdot \exp(a_3 \cdot Br_{class}^{re}) / \exp(a_4 \cdot D_{class})$	0.606	26.07
9	$BR_{class}^{es,T} = a_0 \cdot L_{class} \cdot (Age_{class})^{a_1} \cdot (FI_{class}+1)^{a_2} \cdot \exp(a_3 \cdot Br_{class}^{re}) /(D_{class})^{a_4}$	0.623	27.52
10	$BR_{class}^{es,T} = a_0 \cdot L_{class} \cdot (Age_{class})^{a_1} \cdot (FI_{class}+1)^{a_2} \cdot \exp(a_3 \cdot Br_{class}^{re}) / \exp(a_4 \cdot D_{class})$	0.601	25.67
11	$BR_{class}^{es,T} = a_0 \cdot L_{class} \cdot (Age_{class})^{a_1} \cdot (FI_{class}+1)^{a_2} \cdot (Br_{class}^{re})^{a_3} / \exp(a_4 \cdot D_{class})$	<u>0.575</u>	<u>23.44</u>
12	$BR_{class}^{es,T} = a_0 \cdot L_{class} \cdot (Age_{class})^{a_1} \cdot \exp(a_2 \cdot FI_{class}) \cdot (Br_{class}^{re})^{a_3} /(D_{class})^{a_4}$	0.614	26.78
13	$BR_{class}^{es,T} = a_0 \cdot L_{class} \cdot \exp(a_1 \cdot Age_{class}) \cdot (FI_{class}+1)^{a_2} \cdot \exp(a_3 \cdot Br_{class}^{re}) / \exp(a_4 \cdot D_{class})$	0.696	34.36
14	$BR_{class}^{es,T} = a_0 \cdot L_{class} \cdot \exp(a_1 \cdot Age_{class}) \cdot (FI_{class}+1)^{a_2} \cdot (Br_{class}^{re})^{a_3} / \exp(a_4 \cdot D_{class})$	0.667	31.60
15	$BR_{class}^{as,T} = a_0 \cdot L_{class} \cdot \exp(a_1 \cdot Age_{class}) \cdot (FI_{class}+1)^{a_2} \cdot \exp(a_3 \cdot Br_{class}^{re}) /(D_{class})^{a_4}$	0.703	35.07
16	$BR_{class}^{es,T} = a_0 \cdot L_{class} \cdot \exp(a_1 \cdot Age_{class}) \cdot (FI_{class}+1)^{a_2} \cdot (Br_{class}^{re})^{a_3} /(D_{class})^{a_4}$	0.684	33.24

Table 6.7   Coefficient values in the 11-th equation type (using 80% of the data)

Coefficient	value
a_0	6.85×10^{-4}
a_1	7.2401
a_2	7.8249
a_3	2.8385
a_4	0.0463

Figure 6.1 depicts the estimation and recorded breakage number according to equivalent age aggregation. Through the errors comparison in Figure 6.1 (a), it can be found that the errors of most of the pairs are small, but some of the other errors are a little bigger (e.g., the errors around equivalent age of 15-20 years and 30-40 years). The maximum error is about 16 at the age of 18. In Figure 6.1 (b), most of the errors are small, except for those in the ages of 38, 40, 45 and 47. In Figure 6.2 (a), almost all of the errors are very small. In Figure 6.2 (b), the maximum breakage rate error is 0.00037 breakage/m, which is still a small error. The main reason for some larger errors is that the observation cannot cover its earlier history. There are only 7 years (2003-2009) breakage record in this case study but the service time of most of the pipes are more than 7 years. Because the historical breakage record before observation is not available, the average history record as an input is not accurate. The contrast of errors illustrate that historical data has a great influence to the prediction. Therefore, the formula and efficient must be updated according to new data.

What is of concern is not the single error or part of the errors but the entire available errors as a sample. A proposition of "the differences between estimated and recorded breakage numbers are not significant" can be accepted only after hypothesis testing. In this case study, such a proposition is accepted through *t-test* (students' t test) if significance level is 0.05 (α=0.05). Therefore, the formula is accepted and tested by other independent data.

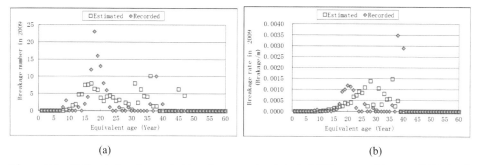

(a)   (b)

Figure 6.1   Estimated and recorded breakage number/rate in 2009, grouped by age (80% data)

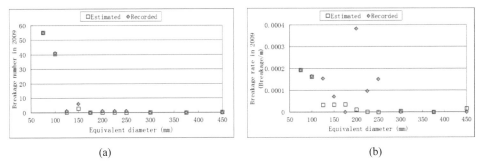

(a) (b)

Figure 6.2   Estimated and recorded breakage number/rate in 2009, grouped by diameter (80% data)

6.2.3 Model Testing

After the formula is fitted, it should be further tested by some other independent asset and failure data. In this study, the remaining 20% of pipes were chosen to test the formula. The feasibility of the method and the prediction errors will be tested. The testing includes:

(1) using the 2003-2005 data to predict breakage number in 2006;

(2) using the 2003-2006 data to predict breakage number in 2007;

(3) using the 2003-2007 data to predict breakage number in 2008;

(4) using the 2003-2008 data to predict breakage number in 2009.

For each of them, hypothesis testing will be applied to test whether the prediction error is significant or not. The hypothesis propositions are "the differences between estimated and recorded breakage numbers are not significant at a certain significance level". If the level is 0.05, all of the above propositions were accepted in this case study. Because the breakage number classified by diameter will be used in cost estimation, they are depicted in Figure 6.3~Figure 6.6.

Through these error comparisons, it can be found that the errors are not so large in most of the predictions. The proposition can be accepted after hypothesis testing. Because of some random factors, more data needed to be applied to calibrate the equation and coefficient. Therefore, using the entire network's data to fitting the equation was the next step.

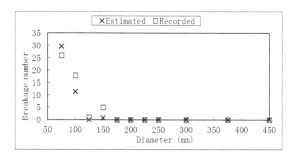

Figure 6.3   Breakage number comparison in 2006 (using 20% data)

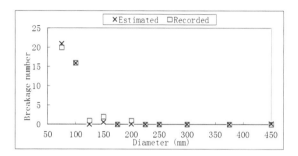

Figure 6.4   Breakage number comparison in 2007 (using 20% data)

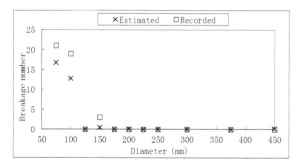

Figure 6.5   Breakage number comparison in 2008 (using 20% data)

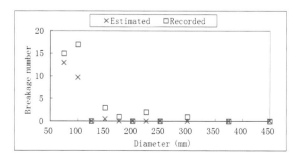

Figure 6.6   Breakage number comparison in 2009 (using 20% data)

6.2.4 Formula Fitting by Using Entire Network's Data

In this step, all of the asset and corresponding data (D, Age, L, FI, historical breakage rate) of 2001-2008 are used to predict breakage number in 2009. A fitting formula of the breakage number in 2009 is produced.

Through classification, aggregation and formula fitting, the formula is as follows:

$$BR_{class}^{es,T} = a_0 \cdot L_{class} \cdot (Age_{class})^{a_1} \cdot (FI_{class} + 1)^{a_2} \cdot (Br_{class}^{re})^{a_3} / \exp(a_4 \cdot D_{class}) \qquad (6.2)$$

Where, $BR_{class}^{es,T}$, L_{class} Age_{class} , FI_{class}, Br_{class}^{re} and D_{class} are estimated equivalent breakage number (in *breaks/year*), equivalent length (in *m*), equivalent pipe age(in *year*), equivalent freezing index (in $°C·d$), recorded breakage rate((in *breaks/year/m*), and equivalent diameter(in *mm*) for the pipe group respectively.

The corresponding error characters are as follows:

Mean square error: 0.708; square sum of residuals: 35.578; related coefficient: 0.9945.

Table 6.8 Coefficient values in the selected equation type (using all of the data)

Coefficients	Values
a_0	9.981×10^{-4}
a_1	7.7620
a_2	7.2431
a_3	3.9758
a_4	0.0492

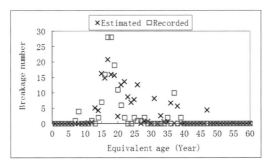

Figure 6.7   Estimated and recorded breakage number in 2009, grouped by age (entire network)

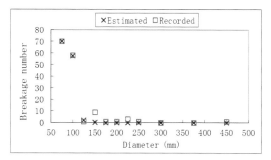

Figure 6.8   Estimated and recorded breakage number in 2009, grouped by diameter (entire network)

Figure 6.7 and Figure 6.8 are the simulation results compared to the recorded breakage number. The same proposition can be accepted though hypothesis testing at a significance level of 0.05. The formula of Eq. (6. 2) with corresponding coefficients (Table 6.8) will be used as future pipe breakage number prediction and current individual pipe condition assessment.

6.2.5 Discussion

1. Error Analysis

The breakage number estimation formula in this research is not always perfect because four limitations still exist. They are complicated deterioration mechanism, complicated external conditions, incomplete records (partly due to short monitoring time horizontal) and inherent randomness. Without the improvement in understanding the deterioration mechanism and longer-term failure record, the accuracy of prediction is difficult to deliver.

Although the model proposed a weighted least squared method to search for the suitable formula according to the existing literature and experience, and the fitness is good for most of the pipe groups in the case study, there are still large errors in some groups. The main reason is that the current formula is a simulation that is mainly based on some experience and assumption instead of theory. Even if the accuracy is improved after modification, the formula type is still based on experience. Such an assumed formula does not involve all the influencing factors and the listed function type cannot either denote completely all the deterioration mechanisms. This is the essential cause of the systematic error. Another reason might be the freezing index, which is derived from historical weather records. The practical winter temperature might be quite different from those in history. Frost has more impact on small pipe's break than large pipes.

In pipe condition assessment, historical break record is involved as representing individual's special feature. The accuracy and completeness of actual record is the key for such a synthesis. If the monitoring horizon is not long enough, the randomness impact may play a dominant role in the nominal break number in the case of break occurrence. Because the monitoring horizon does not cover most part of the pipe's whole life, the information or evidence of failure analysis is limited. In contrast, pipe failure will occur sooner or later only if the observation time is long enough since deterioration exists. Therefore, the error and the randomness in general deterioration tendency's prediction can be reduced only if the monitoring duration is extended.

2.   Understanding of the Fitting Formula

What the decision maker is concerned with is the failure number of different diameters for entire distribution networks because it determines the failure cost. For example, only if the total failure number of the same diameter (e.g., 150 mm) pipes is estimated, the repair and other associated costs can be predicted for these pipes. In the same way, all the other diameter pipe groups' cost can be predicted. Then the entire network's repair cost is known. Because the unit cost is different for different diameter pipes, the failure number must be classified by diameter.

The formula is derived from some pseudo homogenous pipe groups. The equivalent parameters are taken as the common features. Meanwhile, some influencing factors are neglected in classification. The fitting formula can be regarded as the most likely estimated breakage number for such a group.

The estimated failure can be understood at a group of pipes level. Due to the great randomness, the estimated failure number for a group of pipes is the most likely number in estimation. For example, the estimated failure number is one in a specified year for a specified pipe group constituted with a group of exactly same pipes. This means the total failure number expectation of this group is one and all of them have the same likelihood but in which pipe the failure occurs is unknown in the estimation.

The basic intention of aggregation is to reduce some random, unknown or ignored factors' influence and to reinforce the common feature's influence. In the same group, the random factor's influence from an individual pipe to the total burst rate can be weakened with the total pipe length increasing. It is expected that the total service length (i.e. pipe length multiplied by service time span in observation period) in a group is as long as possible.

3.   Individual Pipe's Condition and Overall Network's Failure Estimation

The individual pipe's condition and overall network's failure estimation are correlated closely, but they are different in further application. A decision maker cares for two issues: (1) individual pipe's condition; and (2) the overall network's failure number in a specified year. The former is the foundation to judge a pipe's criticality which is the measurement of rehabilitation priority. The latter is the foundation to estimate the total network's break number and the cost or loss owing to pipe failure. The exact failure location and time is difficult to predict but the total failure number is wanted.

Individual's failure probability estimation is beyond this paper's scope. It needs more individual's features and better understanding of the mechanism. Pipe rehabilitation decision is made at the system level instead of individual pipe level analysis. Pipe failure number prediction is almost the primary foundation of the entire analysis. It is not only used in pipe condition assessment but also in rehabilitation cost analysis. In contrast, pipe condition assessment is the foundation of criticality assessment. The difference between failure prediction and pipe condition assessment has been addressed.

There is great uncertainty for an individual pipe's failure prediction because of unknown mechanisms, ignored influence factors and inherent randomness. However, the estimation will be more stable if the object is a group of homogenous pipes. The main reason is that the relatively homogenous characters strengthen the common factor's influence to failure. The inherent failure discipline can stand out only in the case of large sample quantities because of the randomness. The research focuses on the group or network level of failure prediction rather than individual. To grasp the general tendency more is needed than to capture the randomness. Hence, the network level prediction of pipe deterioration analysis is the real foundation and premise of pipe rehabilitation decision. It is very important to distinguish the differences.

4. Breakage Rate Tendency with Age

According to "bathtub curve" theory, a group of pipe's failure rate is usually experienced in three parts, i.e. early failure, random failure and wear out failure. The failure rate decreases in the early failure period, varies slightly in the random failure period and increases in the wear out failure period. In most literatures, the failure rate increases with age if the diameter is the same and this can be thought as the wear out failure period.

However, the nominal breakage rates, which are the calculation results, in this case study data do not support either the bathtub curve theory or accelerated deterioration tendency. The breakage rates vary up and down greatly and randomly without any clear tendency among all groups. Except for some unaccounted for influencing parameters, the main reason of the fluctuant variation is that the total pipe length in some group is not long enough to eliminate random and unaccounted for factors' influence. Another cause is the length of monitoring time. Because the failure and renewal data only cover eight years, which is really short compared to a pipe's designed service life, the direct record is not complete. Only if the total pipe length in a group is long enough, the general break tendency can play a dominant role. Otherwise, the randomness or casual factors dominate the failure results. The rare failure record further requires longer total pipe length in a group so as to reflect the general tendency.

Another interesting discovery is that the breakage rate seems to decrease with age in this network, but this is not true actually. Bathtub curve indicates general failure rate tendency of a group of pipes. The high failure rate in the early period and relatively lower failure rate in random failures period describe such a seemingly abnormal phenomenon. In distribution system's maintenance record, the failures in a water distribution system usually occur on some relatively new water mains instead of old ones. The main reason is that the prematurely deteriorated assets have been purged from the system so that they are not listed in the present inventory, which has a very short observation period. The residual old pipes are elites with quite good qualities. Another reason is that the amount of the asset (i.e. pipe length) installed more than 40 years ago in the current asset inventory are much less than those installed in the past four decades due to the distribution network's expansion with urbanization. The less length of the old pipe makes the failure record appear more randomly and more scarce. One more reason is most of these old pipes more than 40 years old are large diameter pipes, which have a lower deterioration rate than small diameter pipes.

6.3 Case Study of Pipe Criticality Assessment Model

6.3.1 Background

A smaller and simplified pipe network, which is part of the entire pipe network used in the case study of pipe breakage number prediction model, is applied as the case study of pipe criticality assessment model. Table 6.9 summaries the general features of pipes in this case study. The pipe network with node ID is shown in Figure 6.9. There are 116 pipes and 85 nodes (including two water source nodes). Node 84 and 85 refer to the source nodes with fixed water head and elevations of 45 m and 37 m respectively, while the remaining nodes (Nodes 1 to 83) are demand nodes. The minimum pressure requirement at each demand node is 15 mH$_2$O. The predicted peak daily demands at the nodes in each year in the future can be seen as known condition. The commercially available pipes and their installation unit cost are known as well. The asset and historical repair data are known, including pipe ID, length, diameter, roughness (C-value in Hazen–Williams equation), pipe age, breakage record, node demand, node elevation, unit cost of installation (replacement), relining and repair. All the pipe material is cast iron, same as those in the previous case study.

Table 6.9   Available pipe features

Features	Values
Year of Installation	1972-1999
Diameter	100 mm-450mm
Total length	26.99 Km
Number of pipes	116
Number of breakage records	20

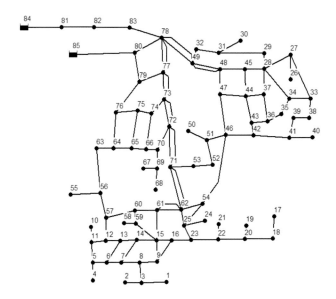

Figure 6.9   Water distribution network

6.3.2 Pipe Criticality Assessment

1.   Pipe Condition Assessment

An individual pipe's condition assessment process is addressed in Section 4.2. The attributes of pipes in this network are known and listed in Table 6.10. Because all of the necessary data, equations and corresponding coefficients are known, the nominal breakage rate for each homogeneous group can be derived. Moreover, the weights can be derived by the length as well. According to the steps in Section 4.2, the nominal breakage rate of 2009 can be obtained.

In this way, each pipes' nominal breakage rate can be calculated so that each pipe's condition can be compared. Through the comparison, it can be found that almost all of the pipes with break record have a higher nominal breakage rate which depicts that the condition of the pipes with break record generally are worse than that of pipes with no breakage record.

Table 6.10 Attributes of Pipes

ID	Node 1	Node 2	L (m)	D (mm)	C-value	Pipe Age (Year)	Historical Breakage rate (br/year/km) (2003-2009)	Nominal breakage rate in 2009 ($\times 10^{-3}$ br/year/km)
1	84	81	100	300	70	38	0	0.442
2	78	80	183	450	110	11	0	0.016
3	17	18	350	100	105	15	0	1.870
4	78	49	160	450	100	19	0	1.763
5	78	49	160	100	70	38	0	0.329
6	49	48	297	400	100	19	0	2.412
7	49	48	297	100	70	38	0	0.329
8	48	47	229	125	105	15	0	1.440
9	47	46	229	100	105	15	0	1.870
10	48	45	183	300	100	19	0	2.007
11	45	44	122	300	105	15	0	0.129
12	44	43	160	100	105	15	0	1.870
13	43	42	104	100	105	15	0	1.870
14	44	47	122	100	105	15	0	1.870
15	43	36	198	100	105	15	0.72	1116.023
16	44	37	122	250	105	15	0	1.222
17	37	28	335	125	105	15	0	1.440
18	37	36	137	200	105	15	0	0.291
19	36	35	213	150	105	15	0	2.344
20	35	34	76	150	105	15	1.88	1590.745
21	34	28	229	150	105	15	0	2.344
22	28	45	480	250	100	15	0	1.222
23	49	32	274	250	105	15	0	1.222
24	32	31	335	250	105	15	0	1.222
25	31	30	228	150	105	15	0	2.344
26	31	29	335	150	105	15	0	2.344
27	29	28	374	100	105	15	0.38	322.636
28	28	27	259	125	105	15	0.55	177.271
29	27	26	350	125	105	15	0	1.440

ID	Node 1	Node 2	L (m)	D (mm)	C-value	Pipe Age (Year)	Historical Breakage rate (br/year/km) (2003-2009)	Nominal breakage rate in 2009 ($\times 10^{-3}$ br/year/km)
30	78	77	312	450	110	11	0	0.016
31	78	77	312	250	70	38	0	0.495
32	77	73	137	450	110	11	0	0.016
33	77	73	137	250	70	38	0	0.495
34	73	72	183	450	110	11	0	0.016
35	73	72	183	250	70	38	0	0.495
36	72	71	46	450	110	11	0	0.016
37	72	71	46	250	70	38	0	0.495
38	71	53	38	250	105	15	0	1.222
39	53	52	213	100	105	15	0.67	922.956
40	52	51	160	200	105	15	0	0.291
41	51	50	167	100	105	15	0	1.870
42	51	46	190	150	105	15	0.75	323.362
43	46	42	152	150	105	15	0	2.344
44	42	41	304	150	105	15	0	2.344
45	41	39	61	150	105	15	0	2.344
46	39	38	61	150	105	15	0	2.344
47	38	33	61	150	105	15	2.34	294.934
48	33	27	274	150	105	15	0	2.344
49	41	40	213	125	105	15	0	1.440
50	34	33	91	100	105	15	0	1.870
51	71	62	548	450	110	11	0	0.016
52	71	62	548	250	70	38	0	0.495
53	62	54	167	250	105	15	0	1.222
54	54	46	472	150	105	15	0	2.344
55	25	54	100	125	100	15	0	1.440
56	25	62	247	450	110	11	0	0.003
57	25	24	320	100	95	15	0	1.870
58	85	80	122	450	95	20	0	0.980
59	80	79	305	300	95	20	0	2.429

ID	Node 1	Node 2	L (m)	D (mm)	C-value	Pipe Age (Year)	Historical Breakage rate (br/year/km) (2003-2009)	Nominal breakage rate in 2009 ($\times 10^{-3}$ br/year/km)
60	79	77	244	150	95	20	0	3.784
61	79	76	76	200	95	20	0	0.614
62	76	75	231	150	95	20	0	3.784
63	75	74	61	100	95	20	0	0.517
64	74	73	107	150	95	19	0	2.693
65	65	75	171	100	95	19	0	8.700
66	66	74	160	100	95	22	0.89	33048.719
67	70	72	53	150	95	22	0	97.390
68	66	70	76	150	95	22	0	97.390
69	66	65	61	150	95	22	0	97.390
70	65	64	365	150	95	22	0	97.390
71	64	76	518	150	95	22	0	97.390
72	64	63	411	125	95	22	0	46.226
73	63	56	305	100	95	22	0.47	8743.367
74	70	69	84	125	95	22	0	46.226
75	69	68	168	100	95	22	0.85	29014.325
76	69	67	198	100	95	22	0.72	19161.014
77	62	61	99	300	110	11	0	0.010
78	62	61	99	200	70	38	0	0.341
79	61	15	701	200	70	38	0	0.341
80	60	61	676	200	110	11	0	0.020
81	60	59	213	150	95	30	0	0.006
82	59	15	411	150	95	30	0.34	2209.200
83	59	58	152	100	95	30	0	0.030
84	57	60	114	150	95	33	0	2.126
85	56	57	518	100	95	33	0	7.626
86	55	56	198	100	95	33	0	7.626
87	57	12	350	150	95	33	0.41	6989.187
88	25	23	708	300	110	11	0	0.010
89	16	23	61	200	100	19	0	4.905

ID	Node 1	Node 2	L (m)	D (mm)	C-value	Pipe Age (Year)	Historical Breakage rate (br/year/km) (2003-2009)	Nominal breakage rate in 2009 ($\times 10^{-3}$ br/year/km)
90	16	15	457	350	110	11	0	0.003
91	15	14	129	300	95	30	0	0.003
92	14	13	129	250	95	33	0	1.429
93	13	12	304	200	95	33	0.47	943.741
94	12	11	76	150	95	33	0	2.126
95	11	10	365	150	95	24	0.39	661.050
96	15	9	137	100	95	24	0	161.800
97	14	7	167	100	95	24	0	161.800
98	6	13	289	100	95	24	0	161.800
99	11	5	309	150	95	24	0	89.293
100	9	16	411	200	100	19	0	4.905
101	9	8	30	250	100	19	0	0.335
102	8	7	69	200	100	19	0	4.905
103	7	6	243	200	100	19	0.59	34.713
104	6	5	236	100	100	19	0	8.700
105	5	4	503	100	100	19	0.28	589.922
106	8	3	137	150	95	19	0	2.693
107	3	1	365	100	95	19	0.39	1126.588
108	2	3	259	100	95	19	0	8.700
109	23	22	259	125	100	19	0	4.076
110	22	20	122	125	100	19	0	4.076
111	20	18	243	125	100	19	0.59	1172.123
112	19	20	373	100	95	19	0.38	1069.613
113	21	22	137	100	95	19	0	8.700
114	81	82	200	300	70	38	0	0.442
115	82	83	300	300	70	38	0	0.442
116	83	78	345	300	70	38	0	0.442

2.   Pipe Significance Assessment

Through the definition of significance index (Eq. (4.7)), the flow rate in each pipe, length, C-value, and diameter are the essential known parameters. Only the flow rates in pipes need to be obtained through hydraulic calculation. The attributes of the pipes is listed in Table 6.10. Each node's elevation and water demand is listed in Table 6.11.

Table 6.11   Elevation and water demand on nodes

Node ID	Elevation (m)	Demand (L/s)	Node ID	Elevation (m)	Demand (L/s)	Node ID	Elevation (m)	Demand (L/s)
1	497.20	11.6	30	513.00	4.8	59	512.70	1.3
2	503.50	4.2	31	511.30	8.8	60	514.60	7.6
3	503.50	1.0	32	511.10	7.9	61	522.10	0.6
4	498.80	3.2	33	509.90	1.2	62	523.20	12.9
5	500.00	8.3	34	509.40	2.6	63	510.50	4.3
6	500.70	8.3	35	510.40	0.0	64	516.50	15.3
7	502.80	2.8	36	511.10	4.8	65	519.50	2.2
8	506.30	0.0	37	513.60	5.6	66	519.50	2.9
9	507.20	3.0	38	509.90	2.6	67	522.70	2.9
10	503.50	14.4	39	511.00	0.9	68	522.70	1.4
11	506.70	0.8	40	512.10	1.0	69	519.10	0.5
12	507.30	14.2	41	512.90	9.5	70	520.80	0.3
13	505.00	1.0	42	513.10	5.8	71	521.40	0.2
14	503.10	1.0	43	514.30	2.9	72	522.00	0.8
15	508.70	22.3	44	515.90	2.2	73	522.40	2.0
16	509.40	0.7	45	513.80	3.4	74	523.20	0.5
17	502.10	2.7	46	515.40	2.4	75	522.90	2.3
18	500.60	2.9	47	517.40	13.8	76	522.40	0.0
19	508.00	1.2	48	515.10	2.8	77	522.10	5.5
20	506.60	0.6	49	520.50	2.6	78	523.60	0.0
21	509.00	4.6	50	518.70	1.0	79	523.60	0.0
22	508.00	2.6	51	518.70	1.3	80	520.40	0.0
23	502.80	6.6	52	522.00	1.1	81	534.00	0.0
24	512.10	6.6	53	520.40	1.0	82	530.00	0.0
25	520.60	3.1	54	522.80	0.2	83	525.00	0.0
26	505.80	3.4	55	510.50	0.8	84	545.00	--
27	506.70	8.3	56	510.50	1.3	85	537.00	--
28	507.50	6.7	57	514.00	0.2			
29	508.30	17.6	58	512.70	1.1			

3. Pipe Criticality Assessment Results

Each pipe's nominal breakage rate in 2009 and its SI (significance index) can be calculated. The weights of nominal breakage rate and significance index obtained through the coefficient of variation method were 0.775 and 0.225 respectively. Meanwhile, the weights obtained through the entropy weighting method were 0.810 and 0.190 respectively. Since there is no preference for weight assignment methods, the average value of these two weighting methods is used as the weight of each criterion. Hence, the weights for pipe's nominal breakage rate and significance index are 0.793 and 0.207 respectively. Through the difference of weights, it can be found that the difference of nominal breakage rate values is greater than that of significance index. Therefore, nominal breakage rate provides more information and has a greater weight as well. Accordingly, pipe condition assessment plays a dominate role in the criticality assessment.

The top 25 critical pipes are listed according to their CI (criticality index) ranks in Table 6.12. Except for the five pipes (the pipe IDs are underlined), the CI ranks of most of the other pipes remain unchanged.

Table 6.12   The top 25 critical pipes

CI Rank	CI (Modified TOPSIS)	CI (TOPSIS)	Nominal Breakage Rate (Normalized and weighted)	SI (Normalized and weighted)	Pipe ID
1	0.976	0.798	0.7926	0.0071	66
2	0.852	0.754	0.6958	0.0019	75
3	0.569	0.542	0.4595	0.0064	76
4	0.270	0.255	0.2097	0.0115	73
5	0.253	0.207	1.680×10^{-7}	0.2074	88
6	0.253	0.210	0.1676	0.0396	87
7	0.231	0.193	3.249×10^{-7}	0.1896	51
8	0.168	0.148	1.055×10^{-5}	0.1380	116
9	0.159	0.141	8.118×10^{-6}	0.1306	79
10	0.146	0.131	1.055×10^{-5}	0.1198	115
11	0.145	0.130	3.249×10^{-7}	0.1191	30
12	0.132	0.120	4.082×10^{-7}	0.1085	80
13	0.123	0.112	3.249×10^{-7}	0.1005	2

CI Rank	CI (Modified TOPSIS)	CI (TOPSIS)	Nominal Breakage Rate (Normalized and weighted)	SI (Normalized and weighted)	Pipe ID
14	0.100	0.073	0.0270	0.0548	107
15	0.097	0.090	1.055×10^{-5}	0.0795	114
16	0.090	0.069	0.0530	0.0207	82
17	0.089	0.086	0.0707	0.0018	47
18	0.088	0.082	1.182×10^{-5}	0.0720	52
19	0.086	0.080	3.249×10^{-7}	0.0701	34
20	0.084	0.078	5.779×10^{-7}	0.0684	6
21	0.081	0.077	2.343×10^{-5}	0.0667	58
22	0.081	0.059	0.0226	0.0438	93
23	0.075	0.057	0.0159	0.0452	95
24	0.074	0.068	0.0023	0.0584	71
25	0.068	0.064	0.0001	0.0554	100

6.4 Case Study of Water Main Optimal Rehabilitation Decision Model

6.4.1 Background

The time span of the whole life is determined by the decision maker. Too long or too short a time is not suitable to meet the requirements of such a concept. A ten-year horizon window is taken into account to act as the whole life of the system in this case study, although the time span of ten years is less than the whole life time span in most literature. The main reasons are as follows:

(1) As a case study (or an example), the time span of ten years is long enough to demonstrate the utility and the process of the method and the influence of current decision. If necessary, the same methodology can be applied to a longer time span, such as decades of years.

(2) Pipe breakage forecasts for a group of pipes in the far future are less accurate and less reliable, although the method is reasonable. This is because the existing measured historical data is not enough. For example, pipe burst prediction formula is not accurate enough for long-term bursts prediction. There is only less than ten years' pipe burst record in this case study; pipe burst prediction formula needs to be revised and tested

continuously by future burst data.

(3) The pipe break randomness dominates the general deterioration tendency on a single pipe. An individual pipe's performances in the far future are difficult to predict. Hence, the performance prediction and estimation on a single pipe are not so reliable.

(4) Generally, more attention is paid to the impact of decisions on the current and near future years, while less consideration is given to the impact of decisions on the far future, when making decisions. The scenarios in the far future have less impact on the current decision. This is because possible problems in the far future need not to be solved immediately. In addition, there are great uncertainties in the development process. It is thought that some uncertainties become less and some possible problems in the far future could be solved gradually in the development, rather than now.

For these above reasons, a relatively short time span (ten years) is applied in this case study.

The pipe deterioration and rehabilitation time step in the research is one year. The annual pipe deterioration estimation will be done and corresponding rehabilitation decision will be generated through optimization. In some research, the analysis time step is a few years (e.g. 5 years). This is mainly because the object in those researches is the entire network's rehabilitation master plan, which focuses on the main pipes, and the total cost is usually the optimal objective. For the rehabilitation detailed plan, the objects include more and smaller pipes than those in a master plan. In practice, pipe renewal action is taken out annually. This result illustrated that timely renewal will reduce cost and improve performance.

Since the unit installation cost is known, the total installation cost for the entire network is also known. In addition, the unit costs are also fixed. Only if pipe failure occurs, repair must be done. Moreover, there are some assumptions to focus on the essential part of this case study:

(1) Each nodal water demand expectation can be predicted through a fixed ratio to the total demand expectation.

(2) Water demand and pipe roughness are accounted for as determined parameters.

(3) Pipe material is the same in case of replacement so that the breaks prediction formula does not change.

(4) The diameter increase in one rehabilitation stage cannot be more than 50 mm if the old

pipe diameter is less than 500 mm, and cannot be more than 100 mm if the old pipe diameter is equal to or above 500 mm. Such an assumption can narrow the option range and save on searching time. Because the old pipe's diameter is known and most water demand increases gradually, this assumption is practical in most cases. If one pipe's diameter changes frequently during a short time, it can be assumed that the pipe diameter should be enlarged to the biggest in the beginning. This adjustment can be done manually.

(5) The minimum pressure requirement for each node and insufficient pressure index requirement are known.

In addition to the attributes of pipes and nodes, some necessary known conditions as follows:

(1) The appropriate annual network's maintenance cost and renewal expenditure is assumed to be 5% for each stage in this case study. In practice, the total available budget depends on decision maker's affordability and service requirements. It is usually known in most water utilities.

(2) Annual budget for network maintenance and renewal does not change greatly. Meanwhile, the budget and expenditure are thought to be almost balanced every year.

(3) This decision is completely acceptable if the total direct cost is less than the limit. Otherwise, the fitness will be modified by the penalty factor.

(4) The requirement of insufficient pressure index is 0.01 in this case study. The insufficient pressure limit is also determined by decision makers according to the service requirement.

Matlab will be used as the computation platform, and EPANET will be applied as the water distribution system's hydraulic computation tool.

6.4.2 Coefficients in Optimal Decision of Present Stage

1. Initial Population

Generally, the early evolution is faster and the latter is slower in the genetic algorithm. In order to make the latter evolution progress faster, the initial population is generated according to the criticality assessment of the pipes and the available budget. Pipes with higher criticality index values are selected for higher probability by a Roulette wheel selection. The available budget is a constraint for the decision. All these efforts are to diversify the population and make them relatively good so as to reduce the computation time.

2. Code String of Chromosome

Each individual is a string with the same length as the pipe number. Each string code (0, 1, 2, or 3) is a gene that represents the rehabilitation action on the pipe. Meanwhile, the total direct cost and each pipe's criticality index are obtained. Thereafter the network's performances (including MIr, estimated total breakage number, water pressure on the nodes) can be calculated.

3. Parameters of Evolution

Because there is no authorized suggested coefficients (e.g., mutation rate and crossover probability) interval for our problem, some trial computation is needed to determine the possible coefficient values. After these coefficient intervals are determined, the real computation begins. The parameters of evolution after trial computation are as follows:

(1) Crossover Probability

Through trial computation, the evolution rate does not change much if the crossover probability varies with the interval [0.8, 1.0]. Because the parents and offspring will be combined in NSGA II, the crossover probability is close to 1.0 without worrying about losing some good individuals in parent's generation.

(2) Mutation Rate

In order to obtain a group of stable populations, the mutation rate is a decreasing function:

$$p_m = \frac{0.1}{k^{0.5}} \qquad\qquad (6.3)$$

Where, p_m is the mutation rate and k is the generation number.

(3) Population Size

Generally, a larger population size is needed for diversity but the computation load increases as well. In trial calculations, population sizes vary from 10 to 100 for a specified generation number. It is found that the fitness of final evolution results (Pareto solutions and average fitness of Pareto solutions) often do not differ very much if the population size is larger than 20. Hence, the population size is 30 in order to balance the computation load and diversity.

(4) Maximum Generation Number

In trial calculations, it is found that the average values of the final population mean fitness changes greatly before the $100^{th} \sim 150^{th}$ generation. It does not change much after the 150th $\sim 200^{th}$ generation. In order to avoid low efficiency of computation, the maximum generation number is set at 300 in this case study.

(5) Fitness

Breakage number fitness: In Eq. (5.10), the coefficient N_{max} =3. Because the historical annual breakage number is around three, it is thought that the maximum acceptable breakage number in a year should be no more than three after rehabilitation.

Modified resilience (MIr) fitness: In Eq. (5.11), $\alpha\%$ is a subjective threshold. For example, $\alpha\%$ =50% in this case study,   which means the capacity to deal with uncertainty is acceptable if the surplus energy is more than half of the total energy requirement. Otherwise, the capacity is not completely acceptable.

(6) Penalty Factor

The formula of Eq. (5.12) provides the basic form of penalty function, but the specific values of the coefficients in the function need to be determined according to the experience and preference of the decision maker.

When b<1.0, the change of b value has a great influence on the value of penalty factor. The smaller the b value, the penalty factor for exceeding budget will decrease rapidly, and the harsher the punishment. This is suitable for the cases where the budget is very sensitive.

When b>1.0, the change of b value has little influence on the change of penalty factor. The change of penalty factor is more relaxed with the change of over budget. This is suitable for the case where exceeding budget is not sensitive.

The greater the a value, the faster the penalty factor drops, and the more severe the penalty, but the effect of a value change is weaker than that of b value.

Penalty factor p_1: in Eq. (5.13), a_1 =5, b_1 =2, the cost limit *Bud* =0.1 in current stage and *Bud* =0.05 in future annual plans. The formula type and coefficients values are determined by comparison. The one which is suitable for decision maker's preference will be chosen.

Penalty factor p_2: in Eq. (5.14), a_2 =1 and b_2 =2, the insufficient pressure limit is H_{in}^* =0.01.

4. Selection

Before crossover, some better individuals will be selected. The basis for selection is the mean fitness, which is the square root of breakage number fitness and Mlr fitness with penalty.

A method to accelerate evolution in the early stage is to set a higher selection pressure. A power function is applied as the pressure selection function. This will enlarge or reduce the difference between individuals. Some individuals with higher fitness have greater possibility of being selected through the Roulette wheel selection. In addition, the modified fitness includes the mean fitness, break fitness and Mlr fitness. These three criteria are applied alternatively as the guidance to select the individuals before crossover. In this case study, the power is 1 before the 100th generations and 0.5 thereafter.

5. Direct Cost

The coefficients in Eq. (5.7), Eq. (5.8) and Eq. (5.9) are derived from practical operation data through regression.

In Eq. (5.7), $a_1 = 4.8453$, $a_2 = 0.6805$

In Eq. (5.8), $a_3 = 49.232$, $a_4 = 0.0024$

In Eq. (5.9), $b_0 = 0.0009$, $b_1 = 0.1657$, $b_2 = 55.571$,

The unit of pipe diameter is mm and that of pipe length is m.

6.4.3 Comparison of Optimization Algorithms

In this case study, three optimization algorithms will be compared. They are multiple-objective GA, NSGA II and modified NSGA II with induced mutation. The goal of comparison is to find a proper algorithm with a rapid evolution rate, diversified and uniform distribution of near Pareto solutions. For this purpose, some trial computations were done.

6.4.3.1 Multiple-objective GA vs. NSGA II

NSGA II is the development of conventional GA. It is generally thought that NSGA II should be better than conventional GA, but this needs to be tested by the case study. Figure 6.10 is the optimization results using two-objective GA by a population of 30 individuals after 300 generations. The dominant population are measured by modified resilience index (*MIr*) fitness and *BR* (breakage number) fitness after modification. The figure depicts the solutions after the final iteration. The blue circles connected by a blue line are the non-inferior solutions (i.e., near Pareto solutions) in the final generation. The red asterisks represent all the solutions in the final generation. Each blue circle or red asterisk represents one rehabilitation decision, which points out that one of three rehabilitation actions (replacement, relining or no action) is done on each of pipe.

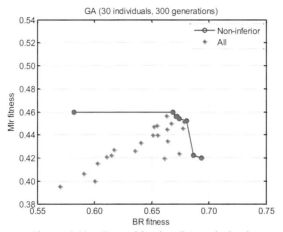

Figure 6.10   Two-objective GA optimization

Figure 6.11 depicts the optimization results in one trial computation using two-objective NSGA II. The *BR* fitness distributes approximately in an interval of 0.64 to 0.77, while the *MIr* fitness is approximate between 0.48 and 0.54. The fitness of the near Pareto solutions in Figure 6.11 is better than that in Figure 6.10. In addition, the solution's distribution scope is also wider than that in Figure 6.10. The blue circles and red asterisks have the same meaning as that in Figure 6.10. NSGA II is more reasonable and better than two-objective GA optimization.

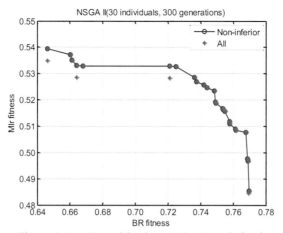

Figure 6.11   Two-objective NSGA II optimization

6.4.3.2 NSGA II vs. Modified NSGA II

Generally, the evolution rate in early generation is relatively fast and slows down gradually. However, the evolution rate in a specified problem is not a definite value. Figure 6.12 shows the populations' average values of mean fitness (the integration of *MIr* fitness and *BR* fitness, Eq.(5.18)) changes with the generation using NSGA II. The average values of mean fitness in a population depict the population's mean fitness level. In this trial computation, the mean fitness after 100 generations is more than 0.62, but the evolution thereafter becomes very slow. Although the evolution rate is not the same in differential computations, the tendency is fast at the early stage and slow at the later stage is general. This is also the general tendency of any generic algorithm.

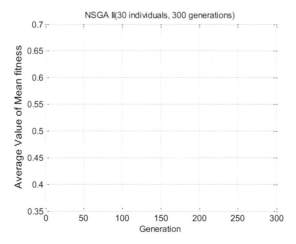

Figure 6.12   Average value of mean fitness changes with generation number

1. Evolution Rate

To improve the optimization rate and find more near Pareto solutions, general NSGA II should be modified. Since flow velocity is an indirect indicator that affects fitness, it is hoped that the pipe with excessive high velocity has more probability to be replaced by a larger pipe. Because this is not a definite selection rule, velocity induced mutation is applied to increase the probability for these pipes to be chosen. Once a pipe is chosen to be replaced because of its high velocity, the new pipe's diameter should also increase.

For the similar reason, the pipe installation/replacement cost can also be used as an induced mutation factor if the solution's total cost exceeds the budget. The pipes to be replaced or relined with expensive cost are prone to be rehabilitated by a cheaper method or even no action at all. Although this affects the performance, the cost is lower and the mean fitness might be improved.

Therefore, velocity and cost can be used as inducing mutation factors to accelerate the evolution rate. The evolution rate of the following four cases will be compared:

(1) NSGA II (without induced mutation);
(2) NSGA II with velocity induced mutation;
(3) NSGA II with cost induced mutation;, and
(4) NSGA II (velocity and cost induced mutation).

In the trial computation, all of the evolution parameters are the same.

The primary indicator to judge an optimization method is closeness with the ideal solution. The mean fitness is such an indicator in this research. Except for mean fitness, the solutions' distribution scope is an auxiliary indicator to judge the advantages and weaknesses of a method. It is hoped that the solutions distribute evenly in a wide range. Standard deviation is employed as such an indicator to measure the solutions' distribution character. If the standard deviation of one fitness is more than that of others methods, such a method is better without considering other indicators.

Figure 6.13 (1)~(3) show the evolution process of the average values of mean fitness of the entire population in each generation. Figure 6.13 (4) zooms in the mean fitness changes of the entire population with the four optimization methods from the 100th to the 300th generation.

Figure 6.14 (1)~(3) show the evolution process of the average values of mean fitness of the Pareto solutions in each generation. Figure 6.14 (4) zooms in the mean fitness changes of Pareto solutions with the four optimization methods from the 100th to the 300th generation. The comparison between NSGA II and other three NSGA II with induced mutation is done one by one. The initial population number is 30 and the total generation number is 300.

Through this comparison, the findings are as follows:

(1) The early stage evolution rate with induced mutation is almost always faster than that without induced mutation, whatever the inducing factor is one of the two or both of them. The mean fitness of Pareto solutions (non-inferior solutions) in early evolution stages might be less than those at the same stages owing to randomness, even though the methods with induced mutation have higher evolution rates in the same stages.

(2) The evolution rate in the later stage (e.g., after 100-150 generations in this case study) becomes slow and the mean fitness differences between the two methods are not significant. These comparisons illustrate that the induced mutation mainly accelerate evolution in the early stages. The reason behind this is that over-emphasis on the cost or velocity will destroy the relatively good gene structure when the population's mean fitness is good enough. Namely, the induced mutation in later stages will bring more negative impact, which is hard to be eliminated.

(3) Through numerous computations, it is found that the three methods with induced mutation are better than, or almost as good as, that without induced mutation in most of cases. One of the trial computation achievements is displayed as representing results of the case study (Figure 6.13 and Figure 6.14), the mean fitness (final evolution results) of the three NSGA II methods with induced mutation are better than those in NSGA II (e.g., velocity induced mutation and velocity plus cost induced mutation), or is as good as those in NSGA II (e.g., cost induced mutation). The reason is that the method with induced mutation has a precise searching capability in the form of probability than the method without induced mutation.

(4) It is difficult to demonstrate which one is better than the other two among the three methods with induced mutation. Through comparison of mean fitness changes in later stages (e.g., from the 200th to the 300th generation in Figure 6.13 (4) and Figure 6.14 (4)), it is found that the distinction of different methods is not obvious. The evolution rate difference among the three induced mutation approaches is not very distinct, although

none of them is worse than the NSGA II without induced mutation. The mean fitness of the final solutions varies between 0.64 and 0.67. Owing to the influence of randomness, the rank of the final solutions' mean fitness might be reversed but the variation range is the same as the interval of [0.64, 0.67]. If viewed from the aspect of standard deviation (Figure 6.15), the differences of *BR* fitness standard deviations are a little obvious in Figure 6.15 (1).   However, the differences of *MIr* fitness standard deviation are not so obvious in Figure 6.15 (2). It must be pointed out that this is not always the case in numerous trial computations. The standard deviations vary more or less in different trials. The difference is not always obvious but the interval of the difference is almost between [0.01, 0.04]. The reason is that the searching is based on probability. The searching process is greatly influenced by randomness in the later fine search. Most of the crossover and mutation have no contribution to the searching in later stages because most of the genes in the achieved solutions are relatively well.

(5) The mean fitness might be going down in the evolution process although the general tendency still is the improvement of mean fitness. For example, the mean fitness of NSGA II with cost-induced mutation fluctuates a little owing to some degradation. A possible reason is that the overall mean fitness might be encumbered by a newly found non-inferior solution which might have better fitness only on one indicator and worse fitness on another indicator. The overall mean fitness might be worse. Such a phenomenon is inevitable but it does not damage the evolution. What we really need is the Pareto solutions (non-inferior solutions) in each evolution step, instead of the solution groups with better mean fitness gradually.

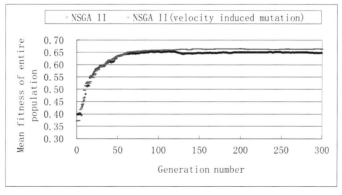

(1)

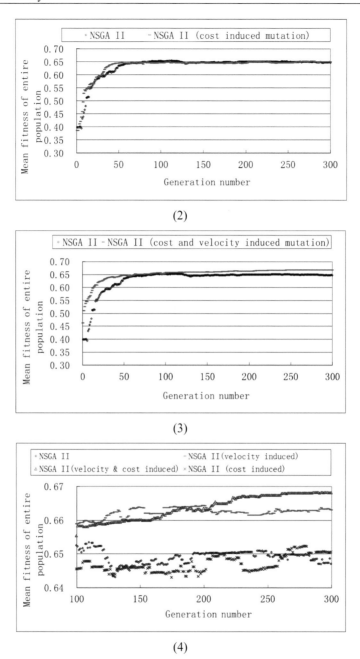

Figure 6.13   Mean fitness changes with generation number (entire population)

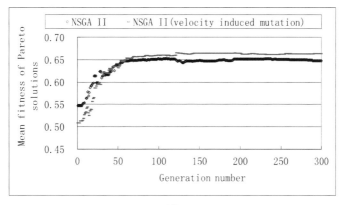

(1)

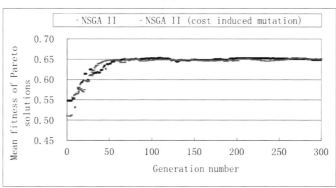

(2)

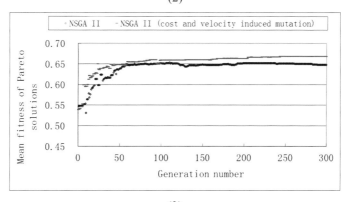

(3)

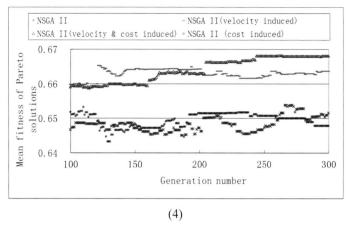

(4)

Figure 6.14   Mean fitness changes with generation number (Pareto solutions)

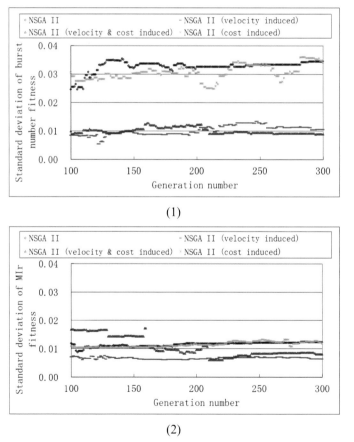

(1)

(2)

Figure 6.15   Standard deviation of *BR* and *MIr* fitness from the 100th to the 300th generation

2.　Fitness in the Final Generation

Finally, near optimal decisions in the last generation, the results from the four approaches are displayed in Figure 6.16. The NSGA II with velocity and cost induced mutation is the best because almost all of the solution's two-dimension fitness is higher than other's. In the aspect of solution's diversity, almost all of the approaches are able to show good performance.

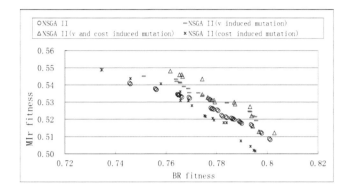

Figure 6.16　Two-dimension fitness of the near Pareto solutions in the last generation

In summary, the induced mutation has a good guidance to accelerate the evolution rate in the early evolution stages, and the fine searching capability is fine in later stages.

3.　Influence of Mutation Rate

An alternative to weaken the negative effects in later evolution is to use a relative higher induced mutation rate in early evolution so as to accelerate the evolution, and a relatively lower induced mutation rate in later evolution so as to create diversified population randomly instead of by some factor's inducement. Although a gradual reduced mutation rate is always applied, the specified mutation rate is still different in the contrast trial calculation. Figure 6.17 is the decision's fitness display of the two approaches in the last generation of one trial computation. In this figure, the triangles are the decisions made with a higher mutation rate in later stages, while the circles are the decisions made with a lower mutation rate in the same stages. The contrast is very clear: the solutions generated by higher mutation rate are distributed in a wider range, but the overall mean fitness of those is not necessarily better.

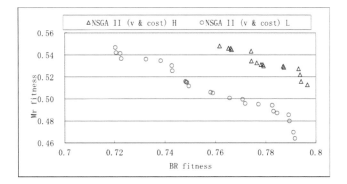

Figure 6.17   NSGA II with velocity and cost induced mutation

Figure 6.18 shows the results with the cost induced mutation. In this figure, the diamonds are the decisions made with a higher mutation rate in later stages, while the squares are the decisions made with a lower mutation rate in the same stage. It can be seen that the mean fitness and the solutions' distribution range do not differentiate too much.

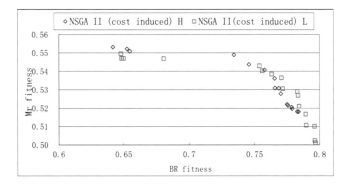

Figure 6.18   NSGA II with cost induced mutation

The accelerate evolution in early stages are significant to reduce the total computation load for a large scale network. The induced mutations in early stages bring a good foundation of widely distributed solutions with relatively good performances. When some relatively good solutions are ready, the further task is to search finely and find more uniformly distributed near Pareto solutions. Because the non-induced mutation and the three induced mutation alternatively lead the mutation operation in early evolution stages, the local searching capability is reinforced. Diversified mutation operation helps avoid being trapped into some local extremes.

4.   Comparison of Intermediate Evolutionary Process

Since the induced mutation has some advantage in the early evolution rate, its impact on solutions' distribution scope and uniformity still needs to be compared. Therefore, the non-inferior solutions in the intermediate evolutionary process are listed in Figure 6.19.

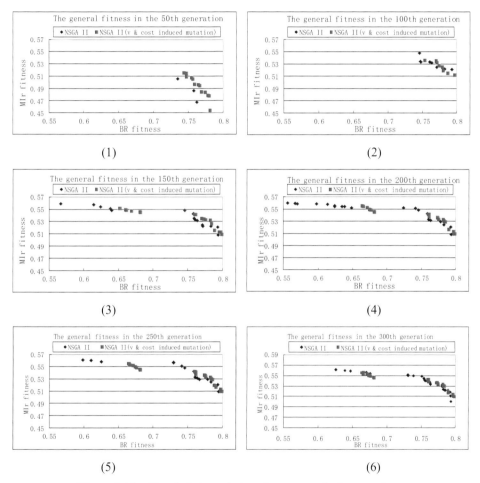

Figure 6.19   Non-inferior solutions in some typical generations

The general evolution tendency is very clear in this series of figures:

(1) Induced mutation is helpful to accelerate the evolution rate in early stages (e.g., before 150th generation in this case study).

(2) The solutions mean fitness changes become small in the later evolution generations, whether the mutation is induced or not.

5. Distribution Scope and Uniformity of Solutions

Except for evolution rate, Pareto solutions' distribution scope and uniformity must be considered as well. In addition, the final solution is also subject to computation load. If too many populations and too many generations are applied, probable better solutions might be achieved at the cost of heavy computation load and long computation time. The computation load could be a burden for a large scale network or a long term planning with multiple planning stages. Therefore, another goal of optimization methodology is to find more near Pareto solutions at the cost of increasing a little computation load. Meanwhile, the diversity of solutions should also be considered.

In order to keep the diversity, population size increase is simple and feasible. After some generation's evolution, the performances of general populations have improved much. It can be thought that all the individuals are to be reserved as parent population, and only some elicits are chosen as a middle generation. .The population size can be doubled in such a method.

If such an idea is applied, the early evolution mainly focuses on evolution rate improvement with induced mutation. The later evolution mainly focuses on near Pareto solutions' fine searching. Lesser selection pressure will be used and population size will be doubled in the middle stage.

Figure 6.19 shows the change process of the solutions' diversity. It can be found that both NSGA II and NSGA II with induced mutation can bring diversified solutions in the evolution process, although the diversity in the beginning of evolution is not so good.

The case study of optimal decision in present stage brings the following findings:

(1) NSGA II is better than single objective GA optimization to deal with multiple optimization problem;

(2) Induced mutation in NSGA II with velocity, or cost, or both as the inducing factor can accelerate the evolution in early stages. This is useful when dealing with large scale networks, in which computation rate is needed more than accurate solutions;

(3) The appropriate mutation rate might help to expand the diversity of the solutions; and

(4) Whether induced mutation is applied or not, the mean fitness of the population is not too distinct in the later evolution stages.

6.4.4 Optimal Decision in Future Stages and Final Optimization Decision

A flow chart of future performance estimation is displayed in Figure 5.4. The initial conditions, premise and current optimal decisions are ready before future decision making.

In this case study, some premise and assumed condition are made before decision making:

(1) The rehabilitation decision making period is one year, which means the rehabilitation job is done every year and is the same as in practice.

(2) The annual budget limit is supposed as 5% of the total assets value.

(3) The allowed insufficient pressure limit is 1% of node's pressure.

(4) The discount rate is 4.5%

(5) The annual water demand increasing rate is 0.5%

(6) The period is 10 years

For each present decision, its performance and corresponding fitness are predicted based on single-objective optimization. In each rehabilitation period, the only optimization objective is the mean fitness (Eq.(5.18)) according to the *BR* fitness and *MIr* fitness with some constraints. The independent variables are rehabilitation decisions on each water main. The computation process and main coefficients are illustrated in Figure 5.5. If one pipe is replaced, its pipe age, roughness, breakage history and even diameter are changed. The pipe breakage number in future stages is predicted through Eq.(6. 2). The *MIr* is obtained through hydraulic computation. Meanwhile, the direct cost due to a specified decision can be calculated if the unit cost of repair and replacement is known. A key step is that each decision in a stage is the premise of the decision in the next stage. Such a rule makes the decision optimized automatically through GA if another decision premise is determined (e.g., the most probable development scenario is taken). After the rehabilitation decision in all of the stages are made, the average fitness of the future stages is the optimization indicator.

In one trial computation, twelve near Pareto solutions for current stage are generated. Because there is no additional evidence to distinguish the performance of the solutions, all of them are chosen for further analysis. In Figure 6.20, variables are mean breakage number fitness and mean *MIr* fitness in the multiple future stages, respectively. Both are modified by the two constraints through penalty factors. The points in this figure depict the current stage solutions' performance in the future. Although these solutions are non-dominated to each other in the current stage, their performances are quite different and not non-dominated any longer. There are two solutions (diamonds with circles in Figure 6.20) still being non-inferior to all the other solutions. Therefore, the decision represented by one of them is chosen as the current rehabilitation decision because its performances are good both in current and multiple future stages. The solution being chosen from these two is to be determined by other conditions and constraints which are not considered in this model.

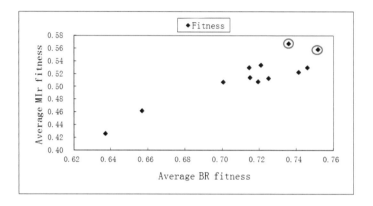

Figure 6.20   Estimated performances in future stages

6.5 Summary

This chapter presents the application of the three models developing the thesis: pipe breakage number prediction model; pipe criticality assessment model; and optimal rehabilitation decision model.

For the pipe breakage number prediction model, coefficients for the model are fitted through weighted non-linear regression. After model testing, the formula is fitted by using the entire network's data. Individual pipe condition can be assessed through a definition formula of nominal breakage rate. The error and the randomness in general deterioration tendency's prediction can be reduced only if the monitoring duration is extended. The nominal breakage rates vary greatly and randomly without any clear tendency among all the groups. The main reason for this is that the total pipe lengths in some of the groups is not long enough to eliminate random and unaccounted factors' of influence. Another reason is the time span of the monitoring is not sufficient.

A sub-zone of the large WDS used in the case study of pipe breakage number prediction model, is applied as the case study of pipe criticality assessment model. In the case study, each pipe's nominal breakage rate and significance index is generated. Through modified TOPSIS, each pipe's criticality index and rank is obtained and this is used to prioritise which pipes are to be decided upon during the optimisation process.

When applying the case studies to the whole-life optimisation model, it was found that NSGA II is more reasonable and better than two-objective GA optimization. Both the solutions' fitness and distribution scope with NSGA II are better than those with two-objective GA. It was also observed that if mutation induction is applied in NSGA II, the early stage evolution rate with induced mutation is almost always faster than that without induced mutation. The evolution rate in later stage becomes slow and the mean fitness differences between the two methods are not significant. The evolution results with induced mutation are better than, or as good as that without induced mutation in most cases. Among the three induced factors (i.e. cost, velocity and both of cost and velocity), it is difficult to demonstrate which one is better. The mean fitness might be fluctuating in the evolution process although the general tendency still is the improvement of mean fitness.

In summary, the induced mutation has a good guidance to accelerate the evolution rate in the early evolution stages, but the fine searching capability is not necessarily very well in later stages. The accelerate evolution in early stages is significant to reduce the total computation load for a large scale network. The induced mutations in early stages bring a good foundation of widely distributed solutions with relatively good performances. When some relatively good solutions are ready, the additional task is to search finely and find more uniformly distributed solutions near the Pareto front. Because the non-induced mutation and three induced mutation alternatively lead the mutation operation in early evolution stages, the local searching capability is reinforced. Diversified mutation operation helps to avoid being trapped into some local extremes.

Chapter 7   Summary, Conclusions and Recommendations

7.1 Introduction

Water distribution systems are a major component of a water utility's asset and may constitute over half of the overall cost of a water supply system. They are critical in delivering water to consumers from a variety of sources. Pipe failures within the distribution system can have a serious impact to both people's daily life and to the wastage of limited, high quality water that has undergone extensive treatment. Hence it is important to maintain the condition and integrity of distribution systems.

The aging and deterioration of pipes in a distribution system are inevitable and this deterioration is responsible for many negative effects to both the water utility and customers. Such effects include an increase in the number and frequency of breaks, increased water loss from the pipelines, reduced hydraulic performance (mainly reduced pressures) and water quality deterioration. Therefore, it is important to be proactive in the rehabilitation of water distribution systems.

Since a water distribution network is a large complex system with interrelated parts, pipe replacement and other maintenance actions will result in far-reaching and complex consequences to performance. Therefore, when undertaking replacement and maintenance, it is important that a systematic analysis is performed to identify the (often counterintuitive) consequences of interventions and to establish the most cost-effective intervention. The replacement and rehabilitation of water mains is an important part of effectively managing a WDS.

This thesis presents a whole-life cost optimisation model for the rehabilitation of water distribution systems. This model allows decision makers to prioritize their rehabilitation strategy in a proactive and cost-effective manner. The optimisation model presented in this thesis, combines a pipe breakage number prediction model and a critically assessment model, that enables the creation of a well-constructed and more tightly constrained optimisation model. This results in improved convergence and reduced computational time and effort. The resulting optimisation model is a multiple-objective one that is solved using an improved genetic algorithm technique. The optimisation model and its related components are demonstrated by applying them to a UK based case study that confirms the feasibility and utility of the developed approach.

7.2 Pipe Breakage Number Prediction Model

Most existing pipe breakage prediction models, whether physically based or statistically based models, are not suitable for optimal replacement and rehabilitation strategies for water distribution systems. The physically based pipe failure prediction models are usually only available for larger and important mains (e.g. backbone mains and water transmission mains) because the data required are usually difficult to obtain. Whereas statistical models are difficult to use in that their expression of uncertainty makes them difficult to incorporate into pipe criticality assessment and optimization decision models.

A new data mining model is proposed in this thesis. Moderate clustering by homogeneous features makes the general deterioration tendency outstanding, which is expressed as a function through non-linear regression. The general deterioration tendency function is derived using readily available data (i.e., pipe material, age, diameter, length, historical breakage record and freezing index) and multiple nonlinear regression. In order to assure the method's reliability, part of the data is employed to fit the formula and the remaining data used to test its validity. The result of the model is the breakage number prediction of a group of pipes (deterioration measure for that group), which can be used for pipe breakage cost estimation in the optimal rehabilitation model.

7.3 Pipe Criticality Assessment Model

The developed pipe criticality assessment model combines an estimate of pipe condition with hydraulic significance to establish a criticality index. A high index value represents those pipes within the network that are of poor condition and are most important for the hydraulic performance of the network. Using this model, a subset of 'important pipes' will established and this subset will then be used by the optimisation model as the group of pipes to focus on when considering replacement and rehabilitation. Clearly, such an approach reduces the search space of the optimisation process and allows the search to be more focussed/targeted. This improves both the solutions generated and the overall computational efficiency of the method.

Pipe structural condition and hydraulic significance are two independent components of the pipe criticality assessment. A pipe's structural condition can be assessed by its nominal breakage rate, which is a virtual concept and a symbolic index that reflects the condition of a pipe structure. A pipe's hydraulic significance reflects the pipes influence on flow and pressure distribution within the network. To combine the two components the criticality model, a multiple-criteria decision making method TOPSIS is used. A modified TOPSIS is developed, that is based on vertical projection distance. After normalization and weighting, the projection of a pipe's condition and significance on an ideal standard axis replace the actual distance in a multi-dimension. This solves the paradox in the conventional TOPSIS method. An objective weighting method is applied to avoid subjective judgment. The ratio of pipe condition and significance after normalization and weighting determine the recommendation for the rehabilitation approach of the pipe.

7.4 Water Main Optimal Rehabilitation Decision Model

A whole life cost optimisation model was developed, where a systematic, long-term view for multi-objective decision making is performed. The primary objective of the optimisation, is to minimize general total costs (i.e. direct, indirect and social costs). The diversified objectives are simplified and integrated so that the selected objectives are more representative and quantified.

Two stages are considered in the optimisation, present stage and future stage.

In the present decision stage, the primary objectives are converted into two measurable objectives and one constraint. Water mains breakage number minimization is one of the main optimization objectives as this is the direct motivation for rehabilitation. This objective is articulated in terms of minimizing indirect and social costs, and maximizing hydraulic serviceability. Another objective is to maximize hydraulic reliability (i.e. modified resilience index) which denotes the network's capability of dealing with uncertain water demand and accidents. Budget limitations (direct costs) are considered as a constraint in optimization model.

For the future decision stage, the optimization objectives and constraints are different from those in the present stage and are more complicated due to uncertainty and chain effects associated with future decision processes. To simplify the process only typical and more relevant scenarios are considered, and two optimization objectives are combined into one. The single objective is the combined fitness of breakage number and modified resilience index with penalty factors. The constraints of the future stage optimisation are the same as for the present stage optimisation. The optimal decisions generated for future stages are used for the following stage's premise. In the model this was regarded as a necessary testing method, rather than a definite decision series as clearly there will be uncertainty associated with future conditions.

To solve the developed optimisation model, a modified NSGA II was developed for the present stage. The two fitness indicators with penalty factors are the two optimization objectives described above. The modified NSGA II involved an induced mutation that acts as a good guidance to accelerate evolution rates, mainly in the early evolution stages where this is required. Moreover, pipe criticality index is used to reduce the search space under consideration by the NSGA II. Whether the induced mutation is applied or not, the mean fitness of the population is not too distinctive in the later evolution stages. Single-objective GA is applied to search for optimal solutions for the future stages under the most probable scenarios. According to the diversified current solutions, their performances, pipe deterioration and most likely corresponding optimal rehabilitation actions in future stages are also estimated. Only the solutions that are non-inferior, both in current stage and future scenarios, are chosen as rehabilitation decisions. This is to guarantee that the decisions are not only the near optimal solution for the present situation but also their potential performances are simulated in the future.

7.5 Recommendation of Future Work

This thesis presents a whole-life cost optimisation model for the rehabilitation of water distribution systems. However, there were limitations to the model, mainly because of numerous objectives, multiple decision stages and the huge computation load. This section provides some recommendations for future work.

(1) The relationship between pipe deterioration and water quality needs further study. Although good pipe condition is a needed to maintain good water quality, relatively little

is known about their relationship. As water quality deterioration is becoming an issue of serious concern, more attention should be paid to this topic. In relation to this, there has been much anecdotal evidence to suggest that pipe condition correlates closely with the ingress of contamination into pipelines, particularly in developing countries where intermittent supply is widespread. The developed models could have good utility in this respect and further study is encouraged.

(2) Due to future global change pressures, there is great uncertainty in relation to the future of both supply and demand of water supply systems. Much discussion has taken place around the need for more flexible systems that have an adaptive capacity, enabling them to respond to uncertainties associated with future changes. It is argued here that the concept of system's flexibility should be embedded into the rehabilitation process. Flexibility is a trade-off between reliability and economic efficiency when dealing with future uncertainty, and further work should be undertaken to consider this.

(3) Data collection and management is a long-term strategy. Data deficiency is still a major barrier for deeper understanding of the deterioration process in pipes. With more data and relevant data the confidence in the predictions generated from deterioration models and the subsequent decisions proposed will be greater. Not only network asset data (including repair and replacement record) but environment data (e.g., temperature and rain) are needed to better understand the process and calibrate models. There also appears to be opportunities of improved data, stemming from the growing interest in the digital economy, internet of things and big data analytics. The implications of these developments on the design and control of water distribution systems and their rehabilitation and replacement approaches will need to be studied.

References

Abo-Sinna, M. A. and A. H. Amer (2005). "Extensions of TOPSIS for multi-objective large-scale nonlinear programming problems." Applied Mathematics and Computation **162**(1): 243-256.

Al-Barqawi, H. and T. Zayed (2006). "Condition Rating Model for Underground Infrastructure Sustainable Water Mains." Journal of Performance of Constructed Facilities **20**: 126.

Al-Zahrani, M., A. Abo-Monasar and R. Sadiq (2016). "Risk-based prioritization of water main failure using fuzzy synthetic evaluation technique." Journal of Water Supply: Research and Technology - AQUA **65**(2): 145-161.

Alvisi, S. and M. Franchini (2009). "Multiobjective Optimization of Rehabilitation and Leakage Detection Scheduling in Water Distribution Systems." Journal of Water Resources Planning and Management-Asce **135**(6): 426-439.

American Water Works Service Co., I. (2002). Deteriorating buried infrastructure management challenges and strategies, American Water Works Service Co., Inc.**:** 1-37.

Arsenio, A. M., I. Pieterse, J. Vreeburg, R. De Bont and L. Rietveld (2013). "Failure mechanisms and condition assessment of PVC push-fit joints in drinking water networks." Journal of Water Supply: Research and Technology - AQUA **62**(2): 78-85.

Arulraj, G. P. and H. S. Rao (1995). "Concept of significance index for maintenance and design of pipe networks." Journal of Hydraulic Engineering **121**(11): 833-837.

Atkinson, S., R. Farmani, F. A. Memon and D. Butler (2014). "Reliability indicators for water distribution system design: Comparison." Journal of Water Resources Planning and Management **140**(2): 160-168.

Bai, H., R. Sadiq, H. Najjaran and B. Rajani (2008). "Condition assessment of buried pipes using hierarchical evidential reasoning model." JOURNAL OF COMPUTING IN CIVIL ENGINEERING **22**(2): 114-122.

Banos, R., J. Reca, J. Martinez, C. Gil and A. L. Marquez (2011). "Resilience Indexes for Water Distribution Network Design: A Performance Analysis Under Demand Uncertainty." Water Resources Management **25**(10): 2351-2366.

Berardi, L., O. Giustolisi, Z. Kapelan and D. A. Savic (2008). "Development of pipe deterioration models for water distribution systems using EPR." Journal of Hydroinformatics **10**(2): 113-126.

Berardi, L., O. Giustolisi and E. Todini (2010). "Accounting for uniformly distributed pipe demand in WDN analysis: enhanced GGA." Urban Water Journal **7**(4): 243-255.

Boxall, J. B., A. O'Hagan, S. Pooladsaz, A. J. Saul and D. M. Unwin (2007). "Estimation of burst rates in water distribution mains." Water Management **160**: 73-82.

Bubtiena, A. M., A. H. El Shafei and O. Jafaar (2012). "Review of rehabilitation strategies for water distribution pipes." Journal of Water Supply: Research and Technology - AQUA **61**(1): 23-31.

Burke, E., G. Kendall, J. Newall, E. Hart, P. Ross and S. Schulenburg (2003). Hyper-heuristics: An emerging direction in modern search technology. Handbook of metaheuristics. F. Glover and G. A. Kochenberger, Springer US: 457-474.

Burke, E. K., M. Gendreau, M. Hyde, G. Kendall, G. Ochoa, E. Ozcan and R. Qu (2013). "Hyper-heuristics: a survey of the state of the art." Journal of the Operational Research Society **64**(12): 1695-1724.

Burn, S., M. Ambrose, M. Moglia, G. Tjandraatmadja and P. Buckland (2004). Management strategies for urban water infrastructure. IWA world water congress & exhibition. Marrakech.

Chen, S. J. J., C. L. Hwang, M. J. Beckmann and W. Krelle (1992). Fuzzy multiple attribute decision making: methods and applications, Springer-Verlag New York, Inc. Secaucus, NJ, USA.

Choi, T. and J. Koo (2015). "A water supply risk assessment model for water distribution network." Desalination and Water Treatment **54**(4-5): 1410-1420.

Chukhin, V., A. Andrianov and V. Orlov (2014). The steel pipe corrosion in drinking water distribution systems and its rehabilitation techniques. 32nd ISTT Annual International No-Dig Conference and Exhibition, October 13, 2014 - October 15, 2014, Madrid, Spain, International Society for Trenchless Technology.

Clair, A. M. S. and S. Sinha (2012). "State-of-the-technology review on water pipe condition, deterioration and failure rate prediction models!" Urban Water Journal **9**(2): 85-112.

Clark, R. M., J. Carson, R. C. Thurnau, R. Krishnan and S. Panguluri (2010). "Condition assessment modeling for distribution systems using shared frailty analysis." Journal American

Water Works Association **102**(7): 81-91.

Clark, R. M., M. Sivaganesan, A. Selvakumar and V. Sethi (2002). "Cost Models for Water Supply Distribution Systems." Journal of Water Resources Planning and Management **128**(5): 312-321.

Clayton, C. R. I., M. Xu, J. T. Whiter, A. Ham and M. Rust (2010). "Stresses in cast-iron pipes due to seasonal shrink-swell of clay soils." Proceedings of the Institution of Civil Engineers-Water Management **163**(3): 157-162.

Constantine, A. G. and J. N. Darroch (1993). Pipeline Reliability: Stochastic Models in Engineering Technology and Management, edited by S. Osaki, DNP Murthy, World Scientific Publishing Co.

Constantine, G., J. Darroch and R. Miller (1996). "Predicting underground pipeline failure." WATER-MELBOURNE THEN ARTARMON- **23**: 9-10.

Cooper, N. R., G. Blakey, C. Sherwin, T. Ta, J. T. Whiter and C. A. Woodward (2000). "The use of GIS to develop a probability-based trunk mains burst risk model." Urban Water **2**(2): 97-103.

Covas, D. and H. Ramos (2010). "Case Studies of Leak Detection and Location in Water Pipe Systems by Inverse Transient Analysis." Journal of Water Resources Planning and Management-Asce **136**(2): 248-257.

Creaco, E., M. Franchini and E. Todini (2016). "The combined use of resilience and loop diameter uniformity as a good indirect measure of network reliability." Urban Water Journal **13**(2): 167-181.

Creaco, E., M. Franchini and T. M. Walski (2016). "Comparison of various phased approaches for the constrained minimum-cost design of water distribution networks." Urban Water Journal **13**(3): 270-283.

Cunha, M. d. C. and J. J. d. O. Sousa (2010). "Robust design of water distribution networks for a proactive risk management." Journal of Water Resources Planning and Management **136**(2): 227-236.

Dandy, G. C. and M. O. Engelhardt (2004). Optimum Rehabilitation of a Water Distribution System Considering Cost and Reliability, ASCE.

Dandy, G. C. and M. O. Engelhardt (2006). "Multi-objective trade-offs between cost and reliability in the replacement of water mains." Journal of Water Resources Planning and

Management-Asce **132**(2): 79-88.

Davies, J. P., B. A. Clarke, J. T. Whiter and R. J. Cunningham (2001). "Factors influencing the structural deterioration and collapse of rigid sewer pipes." Urban Water **3**(1-2): 73-89.

Deb, K., S. Agrawal, A. Pratap and T. Meyarivan (2000). "A Fast Elitism Non-Dominated Sorting Genetic Algorithm for Multi-Objective Optimization: NSGA-II." Proceedings of the Parallel Problem Solving from Nature VI Conference: 849-858.

Del Giudice, G., R. Padulano and D. Siciliano (2016). "Multivariate probability distribution for sewer system vulnerability assessment under data-limited conditions." Water Science and Technology **73**(4): 751-760.

Diao, K., R. Farmani, G. Fu, M. Astaraie-Imani, S. Ward and D. Butler (2014). "Clustering analysis of water distribution systems: identifying critical components and community impacts." Water Science and Technology **70**(11): 1764-1773.

Dridi, L., A. Mailhot, M. Parizeau and J. P. Villeneuve (2009). "Multiobjective Approach for Pipe Replacement Based on Bayesian Inference of Break Model Parameters." Journal of Water Resources Planning and Management-Asce **135**(5): 344-354.

Engelhardt, M., D. Savic, P. Skipworth, A. Cashman, A. Saul and G. Walters (2003). "Whole life costing: Application to water distribution network." Water Science and Technology: Water Supply **3**(1-2): 87-93.

Engelhardt, M. O. (1999). Development of a strategy for the optimum replacement of water mains, PhD thesis, Department of Civil and Environmental Engineering, University of Adelaide, Australia, 1999.

Ennaouri, I. and M. Fuamba (2013). "New integrated condition-assessment model for combined storm-sewer systems." Journal of Water Resources Planning and Management **139**(1): 53-64.

Farley, M. and S. Trow (2003). Losses in water distribution networks: a practitioner's guide to assessment, monitoring and control. London, IWA Publishing.

Farmani, R., G. A. Walters and D. A. Savic (2005). "Trade-off between total cost and reliability for anytown water distribution network." Journal of Water Resources Planning and Management **131**(3): 161-171.

Fox, S., R. Collins and J. Boxall (2016). "Physical investigation into the significance of ground conditions on dynamic leakage behaviour." Journal of Water Supply Research and

Technology-Aqua **65**(2): 103-115.

Francis, R. A., S. D. Guikema and L. Henneman (2014). "Bayesian Belief Networks for predicting drinking water distribution system pipe breaks." Reliability Engineering and System Safety **130**: 1-11.

Fu, G., Z. Kapelan, J. R. Kasprzyk and P. Reed (2013). "Optimal design of water distribution systems using many-objective visual analytics." Journal of Water Resources Planning and Management **139**(6): 624-633.

Fu, G., Z. Kapelan and P. Reed (2012). "Reducing the Complexity of Multiobjective Water Distribution System Optimization through Global Sensitivity Analysis." Journal of Water Resources Planning and Management **138**(3): 196-207.

Fuchs-Hanusch, D., F. Friedl, R. Scheucher, B. Kogseder and D. Muschalla (2013). "Effect of seasonal climatic variance on water main failure frequencies in moderate climate regions." Water Science and Technology-Water Supply **13**(2): 435-446.

García-Cascales, M. S. and M. T. Lamata (2012). "On rank reversal and TOPSIS method." Mathematical and Computer Modelling **56**(5–6): 123-132.

Garcia, D. B. and I. D. Moore (2016). "Evaluation and Application of the Flexural Rigidity of a Reinforced Concrete Pipe." Journal of Pipeline Systems Engineering and Practice **7**(1).

Gheisi, A. R. and G. Naser (2013). "On the significance of maximum number of components failures in reliability analysis of water distribution systems." Urban Water Journal **10**(1): 10-25.

Giustolisi, O. (2010). "Considering actual pipe connections in water distribution network analysis." Journal of Hydraulic Engineering **136**(11): 889-900.

Giustolisi, O. and L. Berardi (2009). "Prioritizing Pipe Replacement: From Multiobjective Genetic Algorithms to Operational Decision Support." Journal of Water Resources Planning and Management-Asce **135**(6): 484-492.

Gong, J. Z., M. F. Lambert, A. R. Simpson and A. C. Zecchin (2013). "Single-Event Leak Detection in Pipeline Using First Three Resonant Responses." Journal of Hydraulic Engineering-Asce **139**(6): 645-655.

Grobler, J., A. P. Engelbrecht, G. Kendall and V. S. S. Yadavalli (2010). Alternative hyper-heuristic strategies for multi-method global optimization. 2010 6th IEEE World Congress on Computational Intelligence, WCCI 2010 - 2010 IEEE Congress on Evolutionary

Computation, CEC 2010, July 18, 2010 - July 23, 2010, Barcelona, Spain, IEEE Computer Society.

Gutiérrez-Pérez, J. A., M. Herrera, R. Pérez-García and E. Ramos-Martínez (2013). "Application of graph-spectral methods in the vulnerability assessment of water supply networks." Mathematical and Computer Modelling **57**(7–8): 1853-1859.

Hadka, D. and P. Reed (2012). "Diagnostic assessment of search controls and failure modes in many-objective evolutionary optimization." Evolutionary Computation **20**(3): 423-452.

Haider, H., R. Sadiq and S. Tesfamariam (2015). "Selecting performance indicators for small and medium sized water utilities: Multi-criteria analysis using ELECTRE method." Urban Water Journal **12**(4): 305-327.

Halhal, D., G. A. Walters, D. Ouazar and D. A. Savic (1997). "Water Network Rehabilitation with Structured Messy Genetic Algorithm." Journal of Water Resources Planning and Management **123**(3): 137-146.

Herrera, M., E. Abraham and I. Stoianov (2016). "A Graph-Theoretic Framework for Assessing the Resilience of Sectorised Water Distribution Networks." Water Resources Management **30**(5): 1685-1699.

Hua, X.-y. and J.-x. Tan (2004). "Revised TOPSIS Method Based on Vertical Projection Distance-Vertical Projection Method [J]." Systems Engineering-theory & Practice **24**(1): 114-119.

Hwang, C. L. and K. Yoon (1981). Multiple attribute decision making: methods and applications: a state-of-the-art survey, Springer Verlag.

Islam, M. S., R. Sadiq, M. J. Rodriguez, H. Najjaran and M. Hoorfar (2014). "Reliability assessment for water supply systems under uncertainties." Journal of Water Resources Planning and Management **140**(4): 468-479.

Jahanshahloo, G. R., F. H. Lotfi and M. Izadikhah (2006). "An algorithmic method to extend TOPSIS for decision-making problems with interval data." Applied Mathematics and Computation **175**(2): 1375-1384.

Jayaram, N. and K. Srinivasan (2008). "Performance-based optimal design and rehabilitation of water distribution networks using life cycle costing." Water Resources Research **44**(1).

Jesson, D. A., H. Mohebbi, J. Farrow, M. J. Mulheron and P. A. Smith (2013). "On the condition assessment of cast iron trunk main: The effect of microstructure and in-service graphitisation on mechanical properties in flexure." **576**: 192-201.

Jin, H. and K. R. Piratla (2016). "A resilience-based prioritization scheme for water main rehabilitation." Journal of Water Supply: Research and Technology - AQUA **65**(4): 307-321.

Kabir, G., S. Tesfamariam, A. Francisque and R. Sadiq (2015). "Evaluating risk of water mains failure using a Bayesian belief network model." European Journal of Operational Research **240**(1): 220-234.

Kapelan, Z., D. A. Savic, G. A. Walters and A. V. Babayan (2006). "Risk- and robustness-based solutions to a multi-objective water distribution system rehabilitation problem under uncertainty." Water Science and Technology **53**(1): 61-75.

Kapelan, Z. S., D. A. Savic and G. A. Walters (2005). "Multiobjective design of water distribution systems under uncertainty." Water Resources Research **41**(11): W11407.

Karamouz, M., K. Yaseri and S. Nazif (2017). "Reliability-based assessment of lifecycle cost of urban water distribution infrastructures." Journal of Infrastructure Systems **23**(2).

Karamouz, M., M. Yousefi, Z. Zahmatkesh and S. Nazif (2012). Development of an algorithm for vulnerability zoning of Water Distribution Network. World Environmental and Water Resources Congress 2012: Crossing Boundaries, May 20, 2012 - May 24, 2012, Albuquerque, NM, United states, American Society of Civil Engineers (ASCE).

Khu, S.-T. and E. Keedwell (2005). "Introducing more choices (flexibility) in the upgrading of water distribution networks: The New York city tunnel network example." Engineering Optimization **37**(3): 291-305.

Kim, J., C. Bae, D. Lee and D. Choi (2012). Statistical estimation model of water pipe deterioration by mathematical function. 12th Annual International Conference on Water Distribution Systems Analysis 2010, WDSA 2010, September 12, 2010 - September 15, 2010, Tucson, AZ, United states, American Society of Civil Engineers (ASCE).

Kim, J. W., G. Choi, J. C. Suh and J. M. Lee (2015). Optimal scheduling of the maintenance and improvement for water main system using Markov decision process. 9th IFAC Symposium on Advanced Control of Chemical Processes, ADCHEM 2015, June 7, 2015 - June 10, 2015, Whistler, BC, Canada.

Kimutai, E., G. Betrie, R. Brander, R. Sadiq and S. Tesfamariam (2015). "Comparison of statistical models for predicting pipe failures: Illustrative example with the city of calgary water main failure." Journal of Pipeline Systems Engineering and Practice **6**(4).

Kleidorfer, M., M. Moderl, F. Tscheikner-Gratl, M. Hammerer, H. Kinzel and W. Rauch (2013). "Integrated planning of rehabilitation strategies for sewers." Water Science and Technology 68(1): 176-183.

Kleiner, Y. (2001). "Scheduling Inspection and Renewal of Large Infrastructure Assets." Journal of Infrastructure Systems 7(4): 136-143.

Kleiner, Y., B. J. Adams and J. S. Rogers (2001). "WATER DISTRIBUTION NETWORK RENEWAL PLANNING." JOURNAL OF COMPUTING IN CIVIL ENGINEERING 5(1): 15-26.

Kleiner, Y. and R. Balvant (2002). "Forecasting Variations and Trends in Water-Main Breaks." Journal of Infrastructure Systems 8(4): 122-131.

Kleiner, Y. and B. Rajani (2001). "Comprehensive review of structural deterioration of water mains: statistical models." Urban Water 3(3): 131-150.

Kleiner, Y., B. Rajani and R. Sadiq (2006a). "Failure risk management of buried infrastructure using fuzzy-based techniques." Journal of Water Supply Research and Technology-Aqua 55(2): 81-94.

Kleiner, Y., B. B. Rajani and S. Wang (2007). Consideration of static and dynamic effects to plan water main renewal. Middle East Water 2007, 4th International Exhibition and Conference for Water Technology. Manama, Bahrain,: 1-13.

Kleiner, Y., R. Sadiq and B. Rajani (2006b). "Modelling the deterioration of buried infrastructure as a fuzzy Markov process." Journal of Water Supply Research and Technology: Aqua 55(2): 67-80.

Kleiner, Y. and B. Rajani (2011). Sampling and condition assessment of ductile iron pipes. World Environmental and Water Resources Congress 2011: Bearing Knowledge for Sustainability, May 22, 2011 - May 26, 2011, Palm Springs, CA, United states, American Society of Civil Engineers (ASCE).

Kleiner, Y. and B. Rajani (2013). "Performance of ductile iron pipes. II: Sampling scheme and inferring the pipe condition." Journal of Infrastructure Systems 19(1): 120-128.

Kutyłowska, M. and H. Hotloś (2014). "Failure analysis of water supply system in the Polish city of Głogów." Engineering Failure Analysis 41: 23-29.

Lansey, K. E. (2000). Optimal design of water distribution systems. Water Distribution System Handbook. New York, McGraw-Hill.

Lee, P. J., H. F. Duan, J. Tuck and M. Ghidaoui (2015). "Numerical and Experimental Study on the Effect of Signal Bandwidth on Pipe Assessment Using Fluid Transients." Journal of Hydraulic Engineering **141**(2).

Li, H. and Q. Zhang (2009). "Multiobjective optimization problems with complicated pareto sets, MOEA/ D and NSGA-II." IEEE Transactions on Evolutionary Computation **13**(2): 284-302.

Li, S., R. Wang, W. Wu, J. Sun and Y. Jing (2015). Non-hydraulic factors analysis of pipe burst in water distribution systems. Computing and Control for the Water Industry. B. Ulanicki, Z. Kapelan and J. Boxall. **119**: 53-62.

Lin, H. Y., B. W. Lin, P. H. Li and J. J. Kao (2015). "The application of the cluster identification method for the detection of leakages in water distribution networks." International Journal of Environmental Science and Technology **12**(8): 2687-2696.

Lippai, I. and L. Wright (2005). Criticality analysis case study: Zone 7 water distribution system. ASCE Pipeline Division Specialty Conference - PIPELINES 2005, August 21, 2005 - August 24, 2005, Houston, TX, United states, American Society of Civil Engineers.

Liserra, T., M. Maglionico, V. Ciriello and V. Di Federico (2014). "Evaluation of Reliability Indicators for WDNs with Demand-Driven and Pressure-Driven Models." Water Resources Management **28**(5): 1201-1217.

Liu, H., D. Savic, Z. Kapelan, M. Zhao, Y. Yuan and H. Zhao (2014). "A diameter-sensitive flow entropy method for reliability consideration in water distribution system design." Water Resources Research **50**(7): 5597-5610.

Liu, Z. and Y. Kleiner (2013). "State of the art review of inspection technologies for condition assessment of water pipes." Measurement: Journal of the International Measurement Confederation **46**(1): 1-15.

Liu, Z. and Y. Kleiner (2014). "Computational Intelligence for Urban Infrastructure Condition Assessment: Water Transmission and Distribution Systems." Ieee Sensors Journal **14**(12): 4122-4133.

Mahinthakumar, G. and M. Sayeed (2005). "Hybrid genetic algorithm - Local search methods for solving groundwater source identification inverse problems." Journal of Water Resources Planning and Management **131**(1): 45-57.

Mahmoodian, M. and C. Q. Li (2016). "Serviceability assessment and sensitivity analysis of cast iron water pipes under time-dependent deterioration using stochastic approaches." Journal of Water Supply: Research and Technology - AQUA **65**(7): 530-540.

Mailhot, A., A. Poulin and J. P. Villeneuve (2003). "Optimal replacement of water pipes." Water Resources Research **39**(5).

Malm, A., O. Ljunggren, O. Bergstedt, T. J. R. Pettersson and G. M. Morrison (2012). "Replacement predictions for drinking water networks through historical data." Water Research **46**(7): 2149-2158.

Marchi, A., E. Salomons, A. Ostfeld, Z. Kapelan, A. R. Simpson, A. C. Zecchin, H. R. Maier, Z. Y. Wu, S. M. Elsayed, Y. Song, T. Walski, C. Stokes, W. Wu, G. C. Dandy, S. Alvisi, E. Creaco, M. Franchini, J. Saldarriaga, D. Paez, D. Hernandez, J. Bohorquez, R. Bent, C. Coffrin, D. Judi, T. McPherson, P. van Hentenryck, J. P. Matos, A. J. Monteiro, N. Matias, D. G. Yoo, H. M. Lee, J. H. Kim, P. L. Iglesias-Rey, F. J. Martinez-Solano, D. Mora-Melia, J. V. Ribelles-Aguilar, M. Guidolin, G. Fu, P. Reed, Q. Wang, H. Liu, K. McClymont, M. Johns, E. Keedwell, V. Kandiah, M. N. Jasper, K. Drake, E. Shafiee, M. A. Barandouzi, A. D. Berglund, D. Brill, G. Mahinthakumar, R. Ranjithan, E. M. Zechman, M. S. Morley, C. Tricarico, G. de Marinis, B. A. Tolson, A. Khedr and M. Asadzadeh (2014). "Battle of the Water Networks II." Journal of Water Resources Planning and Management **140**(7).

Marzouk, M. and A. Osama (2017). "Fuzzy-based methodology for integrated infrastructure asset management." International Journal of Computational Intelligence Systems **10**(1): 745-759.

Mavin, K. (1996). "Predicting the failure performance of individual water mains." Urban Water Research Association of Australia, Research Report **114**.

Misiunas, D. (2005). Failure Monitoring and Asset Condition Assessment in Water Supply Systems. Doctor, Lund University,SWEDEN.

Moglia, M., S. Burn and S. Meddings (2006). "Decision support system for water pipeline renewal prioritisation." Electronic Journal of Information Technology in Construction **11**: 237-256.

Moosavian, N. and B. J. Lence (2017). "Nondominated Sorting Differential Evolution Algorithms for Multiobjective Optimization of Water Distribution Systems." Journal of Water Resources Planning and Management **143**(4): 9.

Mounce, S. R., R. B. Mounce, T. Jackson, J. Austin and J. B. Boxall (2014). "Pattern matching and associative artificial neural networks for water distribution system time series data analysis." Journal of Hydroinformatics **16**(3): 617-632.

Muhammed, K., R. Farmani, K. Behzadian, K. Diao and D. Butler (2017). "Optimal rehabilitation of water distribution systems using a cluster-based technique." Journal of Water Resources Planning and Management **143**(7).

Naderi, M. J. and M. S. Pishvaee (2017). "Robust bi-objective macroscopic municipal water supply network redesign and rehabilitation." Water Resources Management **31**(9): 2689-2711.

Nazif, S., M. Karamouz, M. Yousefi and Z. Zahmatkesh (2013). "Increasing Water Security: An Algorithm to Improve Water Distribution Performance." Water Resources Management **27**(8): 2903-2921.

Nicklow, J., P. Reed, D. Savic, T. Dessalegne, L. Harrell, A. Chan-Hilton, M. Karamouz, B. Minsker, A. Ostfeld, A. Singh and E. Zechman (2010). "State of the art for genetic algorithms and beyond in water resources planning and management." Journal of Water Resources Planning and Management **136**(4): 412-432.

Ofwat. (2003). "Asset Management: A Handbook for Small Water Systems."

Olsson, R. J., Z. Kapelan and D. A. Savic (2009). "Probabilistic building block identification for the optimal design and rehabilitation of water distribution systems." Journal of Hydroinformatics **11**(2): 89-105.

Opila, M. C. and N. Attoh-Okine (2011). "Novel approach in pipe condition scoring." Journal of Pipeline Systems Engineering and Practice **2**(3): 82-90.

Opricovic, S. and G. H. Tzeng (2004). "Compromise solution by MCDM methods: A comparative analysis of VIKOR and TOPSIS." European Journal of Operational Research **156**(2): 445-455.

Osman, H., A. Atef and O. Moselhi (2012). "Optimizing inspection policies for buried municipal pipe infrastructure." Journal of Performance of Constructed Facilities **26**(3): 345-352.

Park, M., H. Lee and B. Sho (2016). "Development of deterioration model of potable water distribution system considering the material properties." International Journal of u- and e-Service, Science and Technology **9**(9): 333-350.

Petit-Boix, A., N. Roige, A. de la Fuente, P. Pujadas, X. Gabarrell, J. Rieradevall and A. Josa (2016). "Integrated Structural Analysis and Life Cycle Assessment of Equivalent Trench-Pipe

Systems for Sewerage." Water Resources Management **30**(3): 1117-1130.

Piratla, K. R. and S. T. Ariaratnam (2011). "Criticality analysis of water distribution pipelines." Journal of Pipeline Systems Engineering and Practice **2**(3): 91-101.

Post, J., I. Pothof, M. C. ten Veldhuis, J. Langeveld and F. Clemens (2016). "Statistical analysis of lateral house connection failure mechanisms." Urban Water Journal **13**(1): 69-80.

Prasad, T. D. and N. S. Park (2004). "Multiobjective genetic algorithms for design of water distribution networks." Journal of Water Resources Planning and Management-Asce **130**(1): 73-82.

Prosser, M., V. Speight and Y. Filion (2015). "Sensitivity analysis of energy use in pipe-replacement planning for a large water-distribution network." Journal of Water Resources Planning and Management **141**(8).

Raad, D. N., A. N. Sinske and J. H. van Vuuren (2010). "Comparison of four reliability surrogate measures for water distribution systems design." Water Resources Research **46**: 11.

Rahmani, F., K. Behzadian and A. Ardeshir (2016). "Rehabilitation of a water distribution system using sequential multiobjective optimization models." Journal of Water Resources Planning and Management **142**(5).

Rajani, B. and Y. Kleiner (2001). "Comprehensive review of structural deterioration of water mains: physically based models." Urban Water **3**(3): 151-164.

Rajani, B. and Y. Kleiner (2002). "Towards Pro-active Rehabilitation Planning of Water Supply Systems." International Conference on Computer Rehabilitation of Water Networks-CARE-W, Dresden, Germany.

Reed, P. M., D. Hadka, J. D. Herman, J. R. Kasprzyk and J. B. Kollat (2013). "Evolutionary multiobjective optimization in water resources: The past, present, and future." Advances in Water Resources **51**: 438-456.

Rogers, P. D. (2011). "Prioritizing water main renewals: Case study of the denver water system." Journal of Pipeline Systems Engineering and Practice **2**(3): 73-81.

Roshani, E. and Y. R. Filion (2014). "Event-based approach to optimize the timing of water main rehabilitation with asset management strategies." Journal of Water Resources Planning and Management **140**(6).

Rowe, R. L., V. Kathula and C. C. Kennedy (2010). "Integrated Conveyance Condition Assessment Techniques Support Asset Management and Capacity Driven Projects." Journal of

Pipeline Systems Engineering and Practice **1**(2): 98-102.

Sadiq, R., B. Rajani and Y. Kleiner (2004). "Probabilistic risk analysis of corrosion associated failures in cast iron water mains." Reliability Engineering & System Safety **86**(1): 1-10.

Saegrov, S., J. F. Melo Baptista, P. Conroy, R. K. Herz, P. LeGauffre, G. Moss, J. E. Oddevald, B. Rajani and M. Schiatti (1999). "Rehabilitation of water networks: Survey of research needs and on-going efforts." Urban Water **1**(1): 15-22.

Salman, B. and O. Salem (2012). "Risk Assessment of Wastewater Collection Lines Using Failure Models and Criticality Ratings." Journal of Pipeline Systems Engineering and Practice **3**(3): 68-76.

Sargaonkar, A., S. Kamble and R. Rao (2013). "Model study for rehabilitation planning of water supply network." Computers, Environment and Urban Systems **39**: 172-181.

Scheidegger, A., J. P. Leitao and L. Scholten (2015). "Statistical failure models for water distribution pipes - A review from a unified perspective." Water Research **83**: 237-247.

Scheidegger, A. and M. Maurer (2012). "Identifying biases in deterioration models using synthetic sewer data." Water Science and Technology **66**(11): 2363-2369.

Scholten, L., A. Scheidegger, P. Reichert, M. Mauer and J. Lienert (2014). "Strategic rehabilitation planning of piped water networks using multi-criteria decision analysis." Water Research **49**: 124-143.

Seica, M. V. and J. A. Packer (2004). "Finite element evaluation of the remaining mechanical strength of deteriorated cast iron pipes." Journal of engineering materials and technology **126**(1): 95-102.

Seifollahi-Aghmiuni, S., O. Bozorg Haddad and M. A. Marino (2013). "Water Distribution Network Risk Analysis Under Simultaneous Consumption and Roughness Uncertainties." Water Resources Management **27**(7): 2595-2610.

Selvakumar, A. and J. C. Matthews (2017). "Demonstration and Evaluation of innovative rehabilitation technologies for nater infrastructure systems." Journal of Pipeline Systems Engineering and Practice **8**(4).

Sempewo, J. I. and L. Kyokaali (2016). Prediction of the Future Condition of a Water Distribution Network Using a Markov Based Approach: A Case Study of Kampala Water. 12th International Conference on Hydroinformatics - Smart Water for the Future, HIC 2016, August 21, 2016 - August 26, 2016, Incheon, Korea, Republic of, Elsevier Ltd.

Shamir, U. and C. Howard (1979). "an Analytical Approach to Scheduling Pipe Replacement." Journal AWWA **71**(5): 248-258.

Sharp, W. W. and T. M. Walski (1988). "Predicting Internal Roughness in Water Mains." Journal American Water Works Association **80**(11): 34-40.

Shih, H.-S., H.-J. Shyur and E. S. Lee (2007). "An extension of TOPSIS for group decision making." Mathematical and Computer Modelling **45**(7-8): 801-813.

Shirzad, A., M. Tabesh and B. Atayikia (2017). "Multiobjective Optimization of Pressure Dependent Dynamic Design for Water Distribution Networks." Water Resources Management **31**(9): 2561-2578.

Sierra, M. R. and C. A. Coello Coello (2005). Improving PSO-based Multi-Objective optimization using crowding, mutation and -dominance. Third International Conference on Evolutionary Multi-Criterion Optimization, EMO 2005, March 9, 2005 - March 11, 2005, Guanajuato, Mexico, Springer Verlag.

Siew, C., T. T. Tanyimboh and A. G. Seyoum (2014). "Assessment of Penalty-Free Multi-Objective Evolutionary Optimization Approach for the Design and Rehabilitation of Water Distribution Systems." Water Resources Management **28**(2): 373-389.

Simpson, A. R., S. Elhay and B. Alexander (2014). "Forest-Core partitioning algorithm for speeding up analysis of water distribution systems." Journal of Water Resources Planning and Management **140**(4): 435-443.

Singh, A. and B. S. Minsker (2008). "Uncertainty-based multiobjective optimization of groundwater remediation design." Water Resources Research **44**(2).

Skipworth, P., M. Engelhardt, A. Cashman, D. Savic, A. Saul and G. Walters (2002). Whole Life Costing for Water Distribution Network Management. London, Thomas Telford.

Skipworth, P. J. (2002). Whole Life Costing for Water Distribution Network Management, Thomas Telford.

Sorge, C., T. Christen and H. J. Malzer (2013). "Maintenance strategy for trunk mains: development and implementation of a high spatial resolution risk-based approach." Water Science and Technology-Water Supply **13**(1): 104-113.

Spiliotis, M. and G. Tsakiris (2011). "Water Distribution System Analysis: Newton-Raphson Method Revisited." Journal of Hydraulic Engineering **137**(8): 852-855.

Srinivas, N. and K. Deb (1994). "Multiobjective Optimization Using Nondominated Sorting in Genetic Algorithms." Evolutionary Computation **2**(3): 221-248.

Stephenson, D. (2005). Water Services Management. London, UK, IWA Publishing.

Tabesh, M. and H. Saber (2012). "A Prioritization Model for Rehabilitation of Water Distribution Networks Using GIS." Water Resources Management **26**(1): 225-241.

Tanyimboh, T. T. and P. Kalungi (2008). "Optimal long-term design, rehabilitation and upgrading of water distribution networks." Engineering Optimization **40**(7): 637-654.

Tanyimboh, T. T., C. Siew, S. Saleh and A. Czajkowska (2016). "Comparison of Surrogate Measures for the Reliability and Redundancy of Water Distribution Systems." Water Resources Management **30**(10): 3535-3552.

Tee, K. F., L. R. Khan, H. P. Chen and A. M. Alani (2014a). "Reliability based life cycle cost optimization for underground pipeline networks." Tunnelling and Underground Space Technology **43**: 32-40.

Tee, K. F., L. R. Khan and H. Li (2014b). "Application of subset simulation in reliability estimation of underground pipelines." Reliability Engineering & System Safety **130**: 125-131.

Todini, E. (2000). "Looped water distribution networks design using a resilience index based heuristic approach." Urban Water **2**(2): 115-122.

Todini, E. and S. Pilati (1988). A gradient method for the analysis of pipe networks. Proc. Computer Applications for Water Supply and Distribution, London, Research Studies Press Ltd.

Torii, A. J. and R. H. Lopez (2012). "Reliability Analysis of Water Distribution Networks Using the Adaptive Response Surface Approach." Journal of Hydraulic Engineering **138**(3): 227-236.

Trifunovic, N. (2006). Introduction to Urban Water Distribution, Taylor & Francis.

Vaabel, J., T. Koppel, L. Ainola and L. Sarv (2014). "Capacity reliability of water distribution systems." Journal of Hydroinformatics **16**(3): 731-741.

Vairavamoorthy, K. and M. Ali (2005). "Pipe Index Vector: A Method to Improve Genetic-Algorithm-Based Pipe Optimization." Journal of Hydraulic Engineering **131**(12): 1117-1125.

Vairavamoorthy, K., Sunil D. Gorantiwar, Jimin Yan, H. M. Galgale, M. A. Mobamed-Mansoor and S. Mobam (2006). Water Safety Plans: Risk Assessment of Contaminant Intrusion Into Water Distribution Systems, Water, Engineering and Development

Centre, Loughborough University.

Vamvakeridou-Lyroudia, L. S., G. A. Walters and D. A. Savic (2005). "Fuzzy multiobjective optimization of water distribution networks." Journal of Water Resources Planning and Management-Asce **131**(6): 467-476.

Van Moffaert, K., M. M. Drugan and A. Nowe (2013). Hypervolume-based multi-objective reinforcement learning. 7th International Conference on Evolutionary Multi-Criterion Optimization, EMO 2013, March 19, 2013 - March 22, 2013, Sheffield, United kingdom, Springer Verlag.

Vilanova, M. R. N., P. Magalhaes Filho and J. A. P. Balestieri (2014). "Performance measurement and indicators for water supply management: Review and international cases." Renewable and Sustainable Energy Reviews **43**: 1-12.

Vrugt, J. A., B. A. Robinson and J. M. Hyman (2009). "Self-adaptive multimethod search for global optimization in real-parameter spaces." IEEE Transactions on Evolutionary Computation **13**(2): 243-259.

Walski, T. (2014). How Does Water Distribution Design Really Work? World Environmental and Water Resources Congress 2014: Water Without Borders, June 1, 2014 - June 5, 2014, Portland, OR, United states, American Society of Civil Engineers (ASCE).

Walski, T. M. and A. Pelliccia (1982). "Economic analysis of water main breaks." Journal American Water Works Association **74**(3): 140-147.

Wang, C.-W., Z.-G. Niu, H. Jia and H.-W. Zhang (2010). "An assessment model of water pipe condition using Bayesian inference." Journal of Zhejiang University: Science A **11**(7): 495-504.

Wang, L., H. Zhang and H. Jia (2012). A leak detection method based on EPANET and genetic algorithm in water distribution systems. Software Engineering and Knowledge Engineering: Theory and Practice: Volume 1, Springer Verlag.

Ward, B., D. Smith, D. Savic, J. Roebuck and J. Collingbourne (2017). "Optimized investment planning for high-volume low-value buried infrastructure assets." Journal of Pipeline Systems Engineering and Practice **8**(3).

Wu, W., H. R. Maier and A. R. Simpson (2013). "Multiobjective optimization of water distribution systems accounting for economic cost, hydraulic reliability, and greenhouse gas emissions." Water Resources Research **49**(3): 1211-1225.

Wu, Z. Y. and A. R. Simpson (2002). "A self-adaptive boundary search genetic algorithm and its application to water distribution systems." Journal of Hydraulic Research **40**(2): 191-203.

Wu, Z. Y. and T. Walski (2004). Self Adaptive Penalty Cost for Optimal Design of Water Distribution Systems. Critical Transitions in Water and Environmental Resources Management Salt Lake City, ASCE.

Xu, Q., Q. W. Chen and W. F. Li (2011). "Application of genetic programming to modeling pipe failures in water distribution systems." Journal of Hydroinformatics **13**(3): 419-428.

Xu, X. (2004). "A note on the subjective and objective integrated approach to determine attribute weights." European Journal of Operational Research **156**(2): 530-532.

Yamijala, S., S. D. Guikema and K. Brumbelow (2009). "Statistical models for the analysis of water distribution system pipe break data." Reliability Engineering & System Safety **94**(2): 282-293.

Yan, J. (2006). Risk Assessment of Condition Intrusion into Water Distribution System. PhD, Loughborough University.

Yan, J. M. and K. Vairavamoorthy (2003). Fuzzy Approach for Pipe Condition Assessment. Proceeding of ASCE International Conference on Pipeline Engineering and Construction, Baltimore, MD, United States, American Society of Civil Engineers.

Yannopoulos, S. and M. Spiliotis (2013). "Water Distribution System Reliability Based on Minimum Cut - Set Approach and the Hydraulic Availability." Water Resources Management **27**(6): 1821-1836.

Yoo, D. G., D. Kang, H. Jun and J. H. Kim (2014). "Rehabilitation priority determination of water pipes based on hydraulic importance." Water (Switzerland) **6**(12): 3864-3887.

Yoo, D. G., J. H. Kim and H. D. Jun (2012). Determination of rehabilitation priority order of subareas in water distribution systems considering the relative importance of pipes. 12th Annual International Conference on Water Distribution Systems Analysis 2010, WDSA 2010, September 12, 2010 - September 15, 2010, Tucson, AZ, United states, American Society of Civil Engineers (ASCE).

Zangenehmadar, Z. and O. Moselhi (2016). "Prioritizing deterioration factors of water pipelines using Delphi method." Measurement: Journal of the International Measurement Confederation **90**: 491-499.

Zitzler, E., M. Laumanns and L. Thiele (2001). SPEA2: Improving the strength Pareto evolutionary algorithm. Evolutionary Methods for Design, Optimization and Control with Applications to Industrial Problems. Athens.

Zitzler, E., L. Thiele, M. Laumanns, C. M. Fonseca and V. G. Da Fonseca (2003). "Performance assessment of multiobjective optimizers: An analysis and review." IEEE Transactions on Evolutionary Computation 7(2): 117-132.

Acknowledgments

I take this opportunity to extend my deepest gratitude to my supervisor, Prof. Kalanithy Vairavamoorthy of UNESCO-IHE/TU Delft and currently the Executive Director of IWA (International Water Association).

He has proficiency for sparking ideas, a sharp eye for detail and super analytical skill. I learned the details and skills required for my research studies from Prof. Vairavamoorthy which have been instrumental in the success of this research study. His continuous support, guidance and constant encouragement throughout my research was invaluable to me. On a personal front, I will never forget the support and assistance Prof. Vairavamoorthy extended to me during my PhD studies. His humour, affability, patience and efficient approach to work made it a pleasure to work with him. This study required contacting some organisations and people for data collection, academic discussion etc., and I sincerely appreciate his encouragement and support of me to get in touch with both the research community and water industry through different channels, which benefited my knowledge with great practical focus. Without his support, this study would never have been successful.

I am indebted to Delft Cluster for providing me financial support in the form of a research studentship, which enabled me to carry out this study. I am grateful to UNESCO-IHE and TU Delft for providing me the valuable opportunity of this PhD research. I am also grateful to the University of Birmingham, UK and the University of South Florida, USA for inviting me as a visiting scholar during my PhD research.

I would like to thank the whole research team, Krishna, Danguang, Seneshaw, Jotham, Jochen, Kebreab and Harrison for their support and friendship. I would also like to express my appreciation to Frank Grimshaw from Severn Trent Water for the data collection. Without the help of all of them it would not have been possible for me to complete this dissertation. I wish to thank my colleagues and the staff at UNESCO-IHE, TU Delft, the University of Birmingham and University of South Florida for their help and support.

My heartfelt thanks also go to my parents, my wife, Weiqing, my daughters, Yundi,and Yunshan, for their continuous support, encouragement, patience and understanding during the course of my Ph.D. study.

Curriculum vitae

Born in Hubei Province, China in September 1975, Yi Zhou began his university studies in 1992 at the Department of Architecture and Civil Engineering in the former Wuhan University of Hydraulic and Electric Engineering (WUHEE), China. Yi graduated in June 1996 and, in the same year, he was employed as a teaching assistant in WUHEE (which joined Wuhan University in 2000). From 1999 to 2002, he carried out his MSc study major in Environment Engineering in Wuhan University. His MSc project was carried out part-time whilst employed as a lecturer in Wuhan University. Supported by a scholarship from the China Scholarship Council (CSC), Yi worked in UNESCO-IHE as a visiting scholar in the Sustainable Urban Infrastructure Systems section chaired by Prof. Kalanithy Vairavamoorthy from July 2006 to January 2007. In June 2007, he became a PhD researcher at the UNESCO-IHE/Delft University of Technology. During his PhD, he followed his supervisor, Prof. Kalanithy Vairavamoorthy, to the University of Birmingham and University of South Florida as a visiting scholar to continue with his research. Currently, Yi is a member of the faculty in the Wuhan University, Wuhan, China.

Publications

Publications in international/national journals and proceedings:

Journal papers:

1. Y. Zhou, K. Vairavamoorthy and M. A. M. Mansoor (2009). "Integration of urban water services." Desalination 248(1-3): 402-409.

2. Zhou Yi, CHEN Yongxiang and LI Xi (2011). "Pressure Driven Water Demand Calculation in Water Supply Network." Engineering Journal of Wuhan University. 44(1): 79-82 (*in Chinese*)

3. ZHOU Yi，CHEN Yong-xiang (2012). "Discussion on Optimization Design Objectives of Water Distribution System." China Water ＆ Wastewater. 28(20): 43-47. (*in Chinese*)

4. ZHOU Yi, YU Ming-hui, CHEN Yong-xiang (2014). "Estimation of sub-catchment width in SWMM." China Water & Wastewater . 30 (22): 61-64. (*in Chinese*)

5. ZHOU Yi, CHEN Yong-xiang (2014). "Application Problems and Countermeasures of Sustainable Urban Rainwater Management Technology in China." China Water & Wastewater. 30 (04): 21-24. (*in Chinese*)

6. Zhou Yi, Yu Minghui, Zhang Yichi, Wu Junzhong, Deng Pinya (2015). "Storm Water flooding process simulation in a nuclear power plant with SWMM." Water and Wastewater. 41 (8): 107-111　(*in Chinese*)

7. Chang Qing, Li Jiang-yun, Zhou Yi (2016). "Sensitivity Analysis of Parameters on Infoworks Model of Urban Stormwater and its Variations with Model Scale." China Rural Water and Hydropower. 57(07): 75-78. (*in Chinese*)

Conference papers:

1. Y. Zhou , K. Vairavamoorthy and F. Grimshaw (2009). Development of a fuzzy based pipe condition assessment model using PROMETHEE, Kansas City, MO, United states, American Society of Civil Engineers.

2. Y. Zhou, K. Vairavamoorthy and F. Grimshaw (2009). A Pipe Condition Assessment Model Using PROMETHEE. Computing and Control in the Water Industry (CCWI) 2009, 'Integrating Water Systems', Sheffield, the UK.

3. Y. Zhou , Y. Shang, J. Li and Q. Tang (2016). Stochastic Long Time Series Rainfall Generation Method. The 2016 International Low Impact Development Conference. Beijing，China. (*in English*)

Printed and bound by CPI Group (UK) Ltd, Croydon, CR0 4YY

21/10/2024

01777112-0014